ENCYCLOPAEDIA OF EMINENT THINKERS

VOLUME-XVII: **MOHAMMAD IQBAL**

Volumes by the same author

The Political Thought of ***Veer Savarkar***
The Political Thought of ***M.A. Jinnah***
The Political Thought of ***Mohammad Iqbal***
The Political Thought of ***K.B. Hedgewar***
The Political Thought of ***M.S. Golwalkar***
The Political Thought of ***Lokmanya Bal Gangadhar Tilak***
The Political Thought of ***Pandit Deendayal Upadhyaya***
The Political Thought of ***Gopal Krishna Gokhle***
The Political Thought of ***Maulana Abul Kalam Azad***
The Political Thought of ***Sir Syed Ahmed Khan***

About the Author

Dr. Jai Narain Sharma (*b.* 1951) is Professor and Chairman of the Department of Gandhian Studies, Panjab University, Chandigarh. He is also Hon. Director of Gandhi Bhawan; member of Board of Studies, Nagpur University, Nagpur and Mahatma Gandhi Kashi Vidyapith, Varanasi; member of Research Degree Committee of the Department of Political Science, H.P. University, Shimla: General Secretary of Indian Soviety of Gandhian Studies; and member on the Presidium of the Alliance for Sarvodaya. He regularly writes for leading newspapers, and has published more than hundred research papers in professional journals of repute. Dr. Sharma has many books to his credit including *"Gandhi's View of Political Power"*, *"Economics of Defence: A Study of SAARC Countries"*, *"Economic Thought of Mahatma Gandhi"*, *"Human Resource Management"*, *"Alternative Economics: Economics of Mahatma Gandhi and Globalisation"*, *"Power, Politics and Corruption: A Gandhian Solution"* and *"Research Methodology: The Discipline and its Dimensions.*

ENCYCLOPAEDIA OF EMINENT THINKERS

VOLUME SEVENTEEN

POLITICAL THOUGHT OF MOHAMMAD IQBAL

Jai Narain Sharma
Professor and Chairman
Department of Gandhian Studies
Panjab University, Chandigarh

CONCEPT PUBLISHING COMPANY, NEW DELHI-110059

Cataloging in Publication Data—DK
Courtesy: D.K. Agencies (P) Ltd. <docinfo@dkagencies.com>

Encyclopaedia of eminent thinkers / K.S. Bharathi.
v. 16-20 cm. — (Gandhian studies and peace research series ; 11)
Vols. 16, 19 by Ashu Pasricha; v. 17-18 & 20 by Jai Narain Sharma.
Includes bibliographical references.
Includes indexes.
ISBN 13: 9788170226840 (set)
ISBN 10: 8170226848 (set)

1. Statesmen—India. 2. Political science—India. 3. India—Politics and government. I. Bharathi, K. S. (Kuduva Swamy), 1956- II. Pasricha, Ashu, 1967- III. Sharma, Jai Narain, 1951- IV. Series: Gandhian studies and peace research series ; 11.

DDC 954.035 22

ISBN -13: 978-81-7022-684-0 (Series)
ISBN -10: 81-7022-684-8 (Series)
ISBN -13: 978-81-8069-497-4 (Vol. XVII)
ISBN -10: 81-8069-497-6 (Vol. XVII)

First Published 2008

Published and Printed by
Ashok Kumar Mittal
Concept Publishing Company
A/15-16, Commercial Block, Mohan Garden
New Delhi-110059 (India)
Phones: 25351460, 25351794
Fax: 091-11-25357103
E-mail: publishing@conceptpub.com

DEDICATED
TO
THE SACRED MEMORY
OF
MY MOTHER
Smt. Vageshwari Sharma

PREFACE

The well known German Orientalist, Professor Annemarie Schimmel of Harvard University, has said of Iqbal: "Nobody would assert that he was a prophet, but we may admit that he has been touched by Gabriel's Wing!"

If we borrow her style of expression, we would say that Iqbal was not a revolutionary in the political sense of the word, but he was definitely touched by the revolutionary spirit of the twentieth century. Like a bee, he imbibed nectar from the flowers and turned it into his own honey. This was the creative process of his poetry and thought.

Iqbal was a multi-dimensional personality who often reflected tendencies that seemed to be contradictory. To isolate one dimension of his personality and thought from the other dimensions is to distort the picture of the total man, the total poet. Yet, at times for a study in depth of a particular facet of his personality that dimension would be isolated. This cannot be escaped.

Iqbal was born in the later part of the nineteenth century and matured in early twentieth century. The complexities of the historical situation and the economic development in our country were such that revolutionaries like Aurobindo preached religious revivalism, and deeply religious souls, like Iqbal, welcomed the winds of change, glorified the revolutions of our age and wrote a new kind of poetry which had never been written before.

Secular minds like Gandhi, in the words of Romain Rolland, introduced into human politics the strongest religious impetus of the last two thousand years and succeeded in stirring millions of people to revolt and shake the foundations of the British Empire. Our task is not only to understand their contradictions but to find the basic truth which is beneficial to modern man. These great personalities do not confront each other, they complement each other.

For Iqbal a thinker and a poet who finds himself at home in the company of Nietzshe and Shopenhauer, Marx and Lenin, Darwin and Einstein, Bhartarihari and Galib, Jinnah and Nehru were a difficult subject. Islam and the Quran are his ideals, yet he has a word of praise for the Geeta. The Prophet Mohammad for him is the most perfect man, yet he respects Sri Krishna, Vishwamitra and Guru Nanak and pays his tribute to them. He takes pride in his Brahmin ancestory to be able to understand the secret of Rumi and Tabriz. Such an eclectic mind is like a mirror in which everyone can see reflected his or her face and read in that face his or her own thoughts. To try and understand him is also a way of paying respect to the great poet and philosopher.

On the political level, I try to understand Iqbal as a poet who made us aware of the stirring of Indian awakening and this includes present day Bharat, Pakistan and Bangladesh. At the same time, he is a poet reflecting the Muslim awakening of Asia and the awakening of entire humanity. His greatest contribution is his philosophy of the Self which envisages man as a continuator of God's creation trying to take the imperfect world to perfection. Like a prophet he tried to awaken all the physical and spiritual potentialities of man to make him aware of his creative power and urges. This led him to the quest for a Perfect Man.

Iqbal demands serious attention because his philosophic poetic genius represents, on the one hand, a continuity of tradition and, on the other, an attempt to integrate it with contemporary world philosophy with an intention to set the tradition moving in an upward direction. Iqbal is unique among the contemporaries of his culture as he combines tradition with modernity with an eye on futurity, which in his peculiar symbolism is named eternity. The concept of the immence of future, as a directing energy or a trend, which operates in the manner of 'Amr' (Command), is one of the leading ideas of Iqbal in his religio-philosophical reflection. In other words, Destiny operates as an immanent factor in the world of man. With this perspective in mind, the present work has been attempted.

Personal acknowledgments are no easy matters where so many have been kind. My special thanks are due to Professor S.L. Malhotra, Professor R.N. Pal, Professor N. Radhakrishnan, Dr. K.S. Bharathi, Dr. R.P. Dwivedi, Dr. A.D. Mishra, Professor B.R. Duggal, Professor Raj Bir Singh Yadav, Professor M.L. Sharma, Dr. Meena Dutta, Dr. Ashu Pasricha and Dr. Rashmi Sharma. Beyond that I can give an all embracing salute to the galaxy of the scholars whose works I have consulted.

I am also grateful to Shri Ashok Kumar Mittal, my publisher, for his consistent zeal in the publication of this volume.

Chandigarh **JAI NARAIN SHARMA**

Encyclopaedia of Eminent Thinkers Series

Vol. I. The Political Thought of **Mahatma Gandhi**/ *K.S. Bharti*
Vol. II. The Political Thought of **Nehru**/ *K.S. Bharti*
Vol. III. The Political Thought of **Rabindra Nath Tagore**/ *K.S. Bharti*
Vol. IV. The Political Thought of **Radhakrishnan**/ *K.S. Bharti*
Vol. V. The Political Thought of **Vinoba**/ *K.S. Bharti*
Vol. VI. The Political Thought of **Aurobindo**/ *K.S. Bharti*
Vol. VII. The Political Thought of **Jayaprakash Narayan**/ *K.S. Bharti*
Vol. VIII. The Political Thought of **Vivekananda**/ *K.S. Bharti*
Vol. IX. The Political Thought of **Ambedkar**/ *K.S. Bharti*
Vol. X. The Political Thought of **M.N. Roy**/ *K.S. Bharti*
Vol. XI. The Political Thought of **Dadabhai Naoroji**/ *Ashu Pasricha*
Vol. XII. The Political Thought of **Veer Savarkar**/ *Jai Narain Sharma*
Vol. XIII. The Political Thought of **M.A. Jinnah**/ *Jai Narain Sharma*
Vol. XIV. The Political Thought of **Tej Bahadur Sapru**/ *Ashu Pasricha*
Vol. XV. The Political Thought of **C. Rajagopalachari**/ *Ashu Pasricha*
Vol. XVI. The Political Thought of **Subhas Chandra Bose**/ *Ashu Pasricha*
Vol. XVII. The Political Thought of **Mohammad Iqbal**/ *Jai Narain Sharma*
Vol. XVIII. The Political Thought of **K.B. Hedgewar**/ *Jai Narain Sharma*
Vol. XIX. The Political Thought of **Bhagat Singh**/ *Ashu Pasricha*
Vol. XX. The Political Thought of **M.S. Golwalkar**/ *Jai Narain Sharma*

CONTENTS

1

LIFE AND ACTIVITES

Mohammad Iqbal, the poet philosopher, was born on November 9, 1877 at Sialkot, and died at the peak of his glory and fame in the early hours of April 21, 1938 at Lahore. Sialkot is a border town on Pakistan side of the Punjab; only a few miles beyond the city on the Indian side begins the land man of Jammu and Kashmir. His grandfather, a Sapru Hindu, embraced Islam. He was the first cousin of Sir Tej Bahadur Sapru.

Shaikh Nur Muhammad, Iqbal's father, was not only endowed with the gift of native intelligence and natural curiosity, but also was complimented with a handsome appearance. He had a ruddy complexion, a platinum white beard, and a pair of penetrating, keen eyes. He was generally respected by his peers because of his religious piety and mystic temperament. He kept the company of the Sufis, from whom he acquired a great deal of mystic knowledge. Out of affection Shaikh Nur Muhammad's friends used to call him an *unparh falsafi* (untutored philosopher).[1]

Iqbal's mother, Imam Bibi, was also a deeply religious woman. As the child of a working class family, she acquired no formal secular education except an elementary knowledge of the Quran, different forms of *ibaddt* (religious duties), and deep consciousness of *iman* (belief) and *ihsan* (right doing). These qualities she firmly instilled in their three daughters and two sons Shaikh Atta Muhammad (the eldest child, born in 1860) and Muhammad Iqbal.[2]

Shaikh Nur Muhammad acquired a reputation as a highly skilled tailor. A local official, Deputy Wazir Ali Bilgrami, had hired him and purchased especially for him a new Singer sewing machine, which was then an object of curiosity. Imam Bibi suspected the legitimacy of the Deputy's income, believing that a large part of it had been derived from un-Islamic and illegal pursuits and thus refused to spend any part of her husband's income which he earned by sewing clothes on his new Singer machine on herself. This eventually compelled Shaikh Nur Muhammad to leave Bilgrami's job. He opened a small millinery of his own, specializing in embroidering the caps of Muslim women's *burqa* (the outer garment concealing their appearance). As a capmaker big demand for his craft grew. He was a success. He eventually hired a few skilled workers to meet the increasing demand. When Shaikh Nur Muhammad grew old, he transferred his shop to his son-in-law, Ghulam Muhammad. Under the latter, the trade soon languished and in a short time the shop closed.

Despite his skill and enterprise as a tailor, Shaikh Nur Muhammad had a hard time supporting his large family. His oldest son, Shaikh Atta Muhammad, did not complete high school. His parents married him to the daughter of a retired soldier of the British Indian Army. After his marriage, his father-in-law secured a position for him in the Army. A few years later he entered the Engineering School at Roorki. Still he graduated from the school and rejoined the Army as an overseer of the Mechanical Engineering Service. Shaikh Atta Muhammad's success raised the social status of the family from a working class to a middle-class position and paved the way for the progress of Iqbal.

Iqbal's childhood and adolescent years in Sialkot were no different than the life of most young children of poor and middle-class families. He was athletically inclined, and used to spend hours in the arena wrestling with his friends. Also, he loved partridges and until the end of his life retained very tender feelings for pigeons. Javid Iqbal, the son of the poet, recalls that during the last years of Iqbal's life when he kept indifferent health "he had a great desire that on the roof of our house he should get a large cage built, full with pigeons, and that his bed should be placed in the middle of them." Iqbal's teacher, Sayyid Mir Hasan, often observed him learning his lessons while tenderly holding a partridge in his hand. Once he asked him: "My dear boy, what peculiar enjoyment do you get out of this?" Iqbal promptly answered: "Please, just hold it and feel".[3]

In the choice of early teachers for Iqbal, his parents also reflected their devotion to Islam. During the period of Iqbal's childhood, four *makatib* in Sialkot (those of Mawlana Ghulam Murtada, Mawlana Abu Abdullah, Mawlana Muhammad Muzammal, and Mawlana Sayyid Mir Hasan) enjoyed a reputation for scholarship and were, even in those days of decline of social structure in general was, reminiscent of the glory of Mughal era. Although the first three concentrated exclusively on Islamic studies and theology, the last one included in syllabus offerings on Arabic and Persian literatures.

Sayyid Mir Hasan (1844-1929), who recognized early Iqbal's poetic talent, was not only an acknowledged scholar of Arabic and Persian literatures, but was also a sympathiser of the Aligarh movement.[4] He offered to tutor Iqbal. Under his tutelege Iqbal graduated from high school (1892) with honours and an award of a scholarship from the Scotch Mission College. Mir Hasan persuaded Shaikh Nur Muhammad to let Iqbal continue his education. Consequently, on May 5, 1893, Iqbal entered the college as a first year student, taking courses in liberal arts.

Iqbal's intellect blossomed at the Scotch Mission College. After attending classes at the college, later in the day Iqbal attended the discourses of Sayyid Mir Hasan at his house. Mir Hasan had committed to memory thousands of Arabic, Persian, and Urdu verses of great masters. He tutored his students, and particularly Iqbal, with an eye to cultivating in them a refined taste and a feeling for Persian and Arabic poetry. He never composed a verse himself, but taught Iqbal the mechanics of classical Urdu and Persian poetry.

Iqbal was now on a road which was destined to bring him success and international fame. However, the year of his high school graduation laid the foundation for the personal anguish and unhappiness which was to mar his life during most of his remaining years. In 1892 his parents married him to Karim Bibi, the daughter of Khan Bahadur Atta Muhammad Khan, an affluent physician in the city of Gujarat. Three children were born to the couple only one of them, a son, survived. Miraj Begum, the eldest child and the exact image of her father, was born in 1895. After a prolonged illness, she died in 1914. The second child, Aftab Iqbal, was born in

1899. Like his father, Aftab studied philosophy, and obtained from the London University a bachelor of arts degree with first class honors in 1923, followed by a master's degree in philosophy two years later. He also qualified as a barrister at law at Lincoln's Inn, London, Iqbal's third male child, who was born in 1901, died soon after birth.

The couple lived together in relative harmony for more than two decades, but differences began to develop and finally became intolerable, especially after Iqbal's return from Europe in 1908. "My life is extremely miserable; they forced my wife upon me," confided Iqbal to a lady friend, Miss Atiya Begum Faizee, in 1909, adding: "I have written to my father that he had no right to arrange my marriage, especially when I had refused to enter into any alliance of that sort. I am quite willing to support her, but I am not prepared to make my life miserable by keeping her with me".[5] Iqbal's affections were largely alienated from Karim Bibi because of his prolonged separations from her and also because of his attachment to Miss Faizee, whom he came to know in Europe. Despite her love and respect for Iqbal, Karim Bibi preferred not to live with a husband who failed to reciprocate her feelings. Finally, in 1916 they decided to live separately, and Iqbal agreed to support her. Karim Bibi survived him. She died on February 28, 1946. Iqbal had supported her until his death in 1938.

In 1895, Iqbal completed his second year at the Scotch Mission College in Sialkot. His teachers recognized his talents and encouraged him to pursue higher studies. The same year, Iqbal went to Lahore and entered the Government College, which was considered to be the best institution of higher learning in the sub-continent. The subjects Iqbal studied for the bachelor of arts degree included Arabic, English literature and philosophy. He graduated *cum laude*, and was also awarded a scholarship for further study, leading toward a master's degree in philosophy. Two years later (1899), he won a gold medal for the unique distinction of being the only candidate who passed the final comprehensive examination.

Iqbal's early years in Lahore opened new vistas of learning and experience. Away from the scrutiny of parents and early teachers, Iqbal tasted an unknown freedom. As a young man, he sowed his wild oats in the streets of Lahore, yet had the good sense to benefit from the best of the city's intellectuals.

At age twenty-two, Iqbal's poetic reputation began making its way into the intellectual. In the last decade of the nineteenth century a *mushaira* (poetical symposium) was regularly organized in the Bazaar Hakiman inside Bhati Gate of the walled city of Lahore. Hakim Amin-ud-Din, a local attorney, was the organiser and initiator, and Hakim Shuja'-ud-Din often presided over the sessions of the *mushaira*. Noted Urdu poets including Mirza Arshad Gorgani and Mir Nazir Husami participated in these sessions. Bhati Gate was then the centre of Lahore's intellectual and cultural activities. Students from local colleges often came, sometimes to enjoy the recitation of poetry, sometimes to compete as budding poets. Iqbal, who was then the poet laureate of the quadrangle, was lured to the poetical symposia of Bhati Gate. In the presence of Gorgani, Iqbal once recited a lyric containing this famous verse:

> Moti samjh kay shan-i karimi nay chun li'ay
> Qatray jo thay mary araq-i infa'l kay.

> His Grace gathered them as pearls, so shining and bright—
> The pearls of perspiration of my remorse.

Mirza Gorgani immediately exclaimed: "Iqbal! Such beautiful verse at your tender age"![16] By the time Iqbal obtained his master's degree in 1899, he was recognized as a promising young poet in the literary circles of Lahore.

In May 1899, a few months after Iqbal's graduation with a master's degree in philosophy, he was appointed Macleod Punjab Reader of Arabic at the University Oriental College of Lahore. His salary was seventy-three rupees a month, and he was required to teach history and economics in addition to Arabic. Probably, Sir Thomas, who was then the acting principal of the Oriental College, was instrumental in his appointment, because Iqbal did not possess the requisite terminal degree in Arabic literature, even though the University Syndicate, in its proceedings of June 23, 1899, had noted that he "stood first in Arabic both at the B.A. and Intermediate examinations".[7] From January 1901 to March 1904, when he resigned from the position of reader, Iqbal taught intermittently as assistant professor of English at Islamia College and at the Government College at Lahore. During this period, he spent no more than a few months of teaching at the Oriental College. As a junior faculty member of the Punjab University, he was not entitled to live in any of the comfortable and spacious houses surrounding the Punjab University and the Government College, as these houses were then reserved exclusively for the British professors.

For Iqbal, the span of time from 1901 to 1905 was filled with hopes and frustrations. The academic profession, which was most congenial to his temperament had lost its luster. The social prestige of a professor ranked lower in the hierarchy of professions. Positions in the civil service and the practice of law offered greater economic and social rewards. Since education was controlled by the British government, freedom of thought and expression was limited, much more so than during the Mughal period when the private academies of eminent scholars could get generous state subsidies without loss of academic freedom.

In order to escape from the academic profession, he studied law simultaneously with his master's course of philosophy. His heart, however, was in philosophy rather than in law; consequently, in December 1898, he failed the preliminary examination in law because he had been indifferent to his study of this subject and as a consequence obtained insufficient points in jurisprudence. Two years later, when he was already Macleod Reader of Arabic at the Oriental College, Iqbal petitioned the chief justice of the Chief Court of the Punjab in Lahore to allow him to re-take the law examination without repeating the schedule of lectures. A great stickler for the rules, Chief Justice Chatterji turned down his petition.[8] Probably this rebuff prompted him to study law later in England.

In 1901, Iqbal tried to enter the civil service of the Punjab and applied to take the examination for the competitive position of extra assistant commissioner (E.A.C.). This ambition also was thwarted when he was removed from the list of the candidates by the medical board.[9]

The denial of an opportunity to compete for a rewarding position was indeed a mental agony. Iqbal must have felt that the built-in mechanism of the socio-political order of colonial India provided no outlet for the fulfilment of his talents. Viewed retrospectively, however, Iqbal's early failures were proverbial blessings in disguise. Had he become an extra assistant commissioner he would have achieved but empty glittering success and spent the remaining years of his life in the wilderness of the rural Punjab as a little czar of the British colonial administration, he would have cheated his destiny.

The end of his frustrations was not yet in sight. In 1903 his elder brother, Shaikh Atta Muhammad, who was then a sub-divisional officer in the Department of Military Works, was dragged into a criminal conspiracy by his rivals and jealous colleagues. In the belief that Shaikh Atta Muhammad would not get a fair trial, Iqbal sent a memorandum to Lord Curzon, the Viceroy of India, describing in detail the facts of his brother's case. Lord Curzon ordered the Baluchistan political agency where Atta Mohammad was posted to enquire into the conspiracy. After thorough investigation Atta Mohammad was exonerated.[10]

The period of his brother's travail, however, provided Iqbal the occasion for notable poetry. Acutely disturbed by the injustice that he perceived would be done to his brother, he gave release to his anguish in a pathetic ode, *Berg-i-Gul*, to Khawja Nizam-ud Din Awliya,[11] praying to the saint to intercede with Allha for His mercy. Khawja Hasan Nizami, the custodian of the saint's shrine and a friend of Iqbal, had the ode sung at the annual URS of 1903. One of the verses of the ode was mounted near the door of the saint's mausoleum.

In 1905, on his way to Europe for advanced graduate studies, Iqbal visited the saint's mausoleum and brought another ode with him, expressing his gratitude for past favours indicating his future aspirations and quest for knowledge. In the stillness of the mausoleum, sitting near the head of the saint's grave, Iqbal recited the ode, *Iltaja i Musafir* (Request of the Traveller).[12]

During his three years of residence in Europe, Iqbal composed twenty-four small poems and lyrics, averaging eight compositions a year. Although poetry had made him famous in India, in Europe he began to doubt the usefulness of his being a poet. "I have made up my mind that I will give up poetry; instead, I would devote my time in the pursuit of something more useful he confided to Shaikh Abdid Qadir, his life long friend and companion for two years in Europe. Qadir reasoned with him to the contrary, saying that his poetry had a magnetic quality capable of inspiring a new life in Muslims of India whose social and economic status was very backward. The final decision was left to Sir Thomas, who fortunately agreed with Qadir.[13] Iqbal bowed to their collective judgment. In 1915, Iqbal published his magnum opus, *Asrar-i Khudi*.[14]

Originally, Iqbal composed 150 verses of *Asrar-i Khudi* in Urdu; subsequently, he destroyed these verses, feeling that he had failed adequately to convey his ideas. Some years later he made another attempt and then forsook the idea of composing *Asrar-i Khudi* in Urdu.

Iqbal studied in both Britain and Germany. In London, he studied at Lincoln's Inn, to qualify at the Bar. And at the Trinity College of Cambridge University, he enrolled as an undergraduate student to earn a bachelor of arts degree. This enrolment was unusual, however, since he already had a master's degree in philosophy from the University of Punjab at Lahore and was simultaneously preparing to submit a doctoral dissertation in philosophy to Munich University. The German university not only allowed him to submit his dissertation in English, but also exempted him from a mandatory stay of two terms on the campus before submitting his dissertation, "The Development of Metaphysics in Persia," to Professor F. Hommel. After his successful defense of the dissertation, Iqbal was awarded the *doctoris philosophiae gradum* on November 4, 1907. The dissertation, which was published the following year in London, was dedicated to T.W. Arnold.[15]

Philosophy being his first love, Iqbal probably wanted to benefit from the lectures of the neo Hegelians, John McTaggart and James Ward, who lectured then to the undergraduates at Cambridge. Moreover, two outstanding orientalists, E.G. Browne and Reynold A. Nicholson, were also at Cambridge lecturing on Persian literature. In view of this, Iqbal's admission as an undergraduate at Cambridge, though unusual, is understandable. Iqbal's intellect was sharpened, and his mental horizon widened under these eminent scholars; they also admired him and recognized his philosophic and poetic talent.[16]

Iqbal was never at home in politics; he was invariably drawn into it. One year after Iqbal's arrival in England, the All-India Muslim League had been established in India. Two years later in London (May 1908) the British Committee of the All-India Muslim League was inaugurated under the presidency of Sayyid Amir Ali, a former judge of the Calcutta High Court and the author of several scholarly studies on Islamic law and history, including *The Spirit of Islam*[17] and *A History of the Saracens*.[18] Even though Iqbal was to depart from England three months later, he was elected to the executive committee and, along with Sayyid Hasan Bilgrami and Sayyid Amir Ali, Iqbal was nominated to a sub-committee to draft the rules and regulations of the British Committee of the Muslim League. With the exception of one brief interruption, Iqbal maintained this relationship with the All-India Muslim League throughout his life.

As to Iqbal's personal life in Europe, very little authentic information is available. Miss Atiya Begum Faizee, who knew Iqbal in Europe, has partially filled the gap.[19]

In 1909 Iqbal's family arranged a second marriage with Sardar Begum, a charming but physically frail young lady belonging to a respectable Kashmiri family from the neighbourhood of Mochi Gate in Lahore. [Here sources disagree, some maintaining that the *nikah* (a ceremony solemnizing the contract of marriage, often without the consummation of the marriage, however) was actually performed, others asserting that only the engagement ceremony was celebrated.] It is established, nevertheless, that the bride did not come to the house of the groom. Iqbal had received anonymous letters impugning the character of the young bride. (Actually this was the mischief of a local lawyer, Nabi Bakhsh, who wanted very much to marry his own son to Sardar Begum.) This situation caused no small amount of misery to Iqbal, but it seems that he did not act prudently in handling this matter. Without terminating his formal matrimonial relationship with Sardar Begum, Iqbal married another girl, Mukhtar Begum, the niece of a very wealthy businessman of Ludhiana. She died in 1924.[20]

A few years after the uncompleted marriage, Sardar Begum wrote a letter to Iqbal chiding him for his injustice and lack of compassion. "I was married to you, a second marriage is now inconceivable to me," Sardar Begum wrote, adding: "I will remain in my present state till I die, and on the day of judgment will hold you (responsible for ruining my life)."[21] Iqbal was married to Sardar Begum for the second time in 1913. She gave Iqbal love, devotion, and peace of mind. However, she died prematurely on May 23, 1935, when she was only thirty-seven years old, and left Iqbal with a son, Javid Iqbal (born in 1924), and a daughter, Mumirah (born in 1930).

When Iqbal came back from Europe in 1908, he started his professional career as an attorney, college professor and poet, all at once. At length, however, the true poet won out at the expense of the professor and the attorney, but not before years of personal turmoil. The yoking of the professor to the poet might have worked if Iqbal had been able to teach in a private university or college like Harvard or Amherst. But the position of assistant professor at the Government College, Lahore promised him only a modest monthly income without guaranteeing academic freedom as it was understood at Cambridge and Oxford or Yale and Princeton. In 1909, Aligarh Muslim University offered him a professorship of history, but he declined the invitation. After deciding to resign his temporary teaching position at the college, Iqbal confided (July 17, 1909) to Miss Faizee: "The Lieutenant Governor was willing to recommend me to the Secretary of State for India for the vacant Professorship in the Lahore Government College, but I gave up the idea of standing as candidate, much against my personal inclination. Force of circumstances compels me to consider things from [the] financial point of view—a point of view which was revolting to me a few years ago. I have decided to continue the legal profession trusting in God's help".[22]

Iqbal so repeatedly versified the proverb "God helps those who help themselves" that today every school boy in Pakistan knows it by heart. Yet he did precious little to help his own law practice, which started in October 1908, and virtually ended in 1934. After ten years of law practice he wrote to his father in 1918: "I have made the determination not to personally benefit from the income derived from the sale of my prose and poetry. Because [poetry] is a God given talent, involving no physical labour, and it should be exercised in the service of mankind, but I have been compelled by need to act otherwise".[23] According to his friends and associates, Iqbal hardly ever earned more than a thousand rupees a month, the income of a moderately successful attorney of the lower courts in today's Pakistan.

Iqbal never really aspired to earn more than he actually needed. Performing the role of an appellate lawyer, he confined his practice to the High Court of judicature at Lahore. It is estimated that including the house and office rent and the salaries of his clerk and domestic servants, his fixed expenses amounted to approximately seven hundred rupees. He would be quite content when he earned enough to cover his overhead, plus a few hundred extra for incidental expenses. Consequently, any additional cases coming to him after the tenth of each month were refused, or were referred to other attorneys; if the client insisted on retaining him he was advised to come before the tenth of the following month.[24]

Iqbal devoted his time primarily to his legal practice and cultural activities, but when elections were introduced in Punjab under the Montagu Chelmsford Reforms, his friends and admirers prevailed upon him to offer himself as a candidate. In November 1926 Iqbal contested a seat from a Muslim district of Lahore. Out of the total of 12,000 registered voters, 68 per cent actually voted; Iqbal obtained 5,675, and his opponent, Malik Muhammad-Din, received 2,498.[25] True to his sedentary habits, Iqbal hardly ever ventured to campaign outside of his residence. His popularity had ensured his success. When the result was declared. Iqbal's volunteers and campaigners rushed to his home and repeatedly shouted, *Iqbal zindabad*, "Long live Iqbal."

In December 1928, Iqbal travelled to South India in order to deliver six lectures on Islam at the request of the Madras Muslim Association. In these addresses, Iqbal declared that he was attempting to "reconstruct Muslim religious philosophy with due regard to the philosophical tradition of Islam and the more recent development in the domains of human knowledge".[26] Having taken three years to compose these lectures, Iqbal viewed them as indicative of his mature philosophic and rational approach to Islam.

Twenty-three years after he returned from Europe in 1908, Iqbal had another occasion to visit Britain and other continental countries. The opportunity was provided by the second (September 7 to December 1, 1931) and the third (November 17 to December 24, 1932) London Round Table Conferences. (Iqbal had not been invited to the first conference.) The Round Table Conferences were called by the British Government to consult with Indian leaders on the problems of constitutional reforms.

Although he remained in Europe for less than a year, European diplomacy and the political and intellectual trends which engulfed Europe in the bloodbath of the late 1930s left a deep impression on Iqbal's mind. (Most of these impressions are reflected in the verses of *Darb-i Kalim* [The Rod of Moses], first published in July 1936.[27] During the second conference Iqbal was occupied with the business of the conference and had no significant opportunity to travel in Europe. In the third conference (1932) his participation was minimal. He spent four months travelling in Europe and meeting scholars and political leaders, including Henri Bergson and Louis Massignon of France and Benito Mussolini of Italy.

In February 1933, Iqbal was back in Lahore. Seven months later Muhammad Nadir Shah, the King of Afghanistan, invited him along with Sayyid Sulaiman Nadwi and Sir Sayyid Ross Masud to visit Kabul. The Afghan king asked them to advise his government in the establishment of a new university and in utilizing the best of modern western and traditional Islamic values in the reorganization of higher education.

Not much is known about the educational recommendations of Iqbal and his associates. On October 10, 1933 he issued a statement saying that "an educated Afghanistan will be the best friend of India".[28] He decried secular education for Afghanistan, because the "complete secularization of education has not produced good results anywhere, especially in Muslim lands." Despite Afghanistan's general backwardness and the conservatism of the royal family, Iqbal complimented the Afghan regime in a statement of November 6: "All the ministers are discharging their duties. Even the orthodox party stands solidly behind these workers and consequently there is, as was stated in our presence by a leading Afghan divine no difference between the *mullas* and the young men in the Afghanistan of today".[29]

After his return from Afghanistan, Iqbal's health steadily deteriorated. However, his intellect remained sharp, and during this time, he conceived many new projects, including proposed studies on Islamic jurisprudence and *Aids to the Study of Quran*. An interpretive study, *Islam As I Understand It*, was to reconcile modern science and philosophy with Islam. In the style of thus *Spake Zarathustra,* Iqbal planned to write *The Book of an Unknown Prophet* discussing metaphysical problems. Iqbal spent considerable mental effort in determining the exact literary style for his bible. None of these studies saw the light of day; political preoccupations and indifferent health made their execution impossible.[30]

With the start of 1938, Iqbal's health sharply deteriorated. Overwhelmed by asthmatic attacks, he steadily grew weaker, but his mind suffered no disability. He died on April 21, 1938.

REFERENCES

1. Malik, Hafiz (ed.), *Iqbal Poet Philosopher of Pakistan,* New York: Columbia University Press, 1971, p. 4.
2. Salik, Abdul Majid, *Dhiker-i-Iqbal*, Lahore, Bazam-i-Iqbal, 1955, p. 48.
3. *Ibid.*, p. 12.
4. Malik, Hafiz (ed.), *Iqbal Poet Philosopher of Pakistan, op. cit.*, p. 8.
5. Faizee, Atiya Begum, *Iqbal*, Bombay, Academy of Iqbal, 1947, pp. 110-112.
6. Salik, Abdul Majid, *Dhiker-i-Iqbal, op. cit.*, p. 55.
7. Malik, Hafiz (ed.), *Iqbal Poet Philosopher of Pakistan, op. cit.*, p. 13.
8. Salik, Abdul Majid, *Dhiker-i-Iqbal, op. cit.*, p. 65.
9. *Ibid.*
10. Hussain Riaz, *Iqbal Poet and His Politics,* New Delhi: Uppal Publishing House, 1987, p. 4.
11. Malik, Hafiz (ed.), *Iqbal Poet Philosopher of Pakistan, op. cit.*, p. 15.
12. *Ibid.*
13. *Ibid.*, p. 17.
14. Iqbal, Mohammad *Asrar-i-Khudi*, Lahore, Ghulam Ali, 1957 (Translated) as '*Secrets of Self*', New Delhi, Arnold-Heinemann, 1978.
15. Hussain Riaz, *Iqbal Poet and His Politics, op. cit.*, p. 15.

16. *Ibid.*
17. Ali, Sayyid Amir, *The Spirit of Islam*, London, Christopher, 1922.
18. Ali, Sayyid Amir, *A History of the Saracens*, London, Macmillan, 1899.
19. Faizee, Atiya Begum, *Iqbal, op. cit.*, pp. 120-122.
20. Malik, Hafiz (ed.), *Iqbal Poet Philosopher of Pakistan, op. cit.*, p. 24.
21. Salik, Abdul Majid, *Dhiker-i-Iqbal, op. cit.*, p. 69.
22. Faizee, Atiya Begum, *Iqbal, op. cit.*, pp. 124-125.
23. Niyazi, Sayyid Nazir, *Maktubat-i-Iqbal*, Karachi, Iqbal Academy, 1957, p. 266.
24. Wahid-u-Din, Syed Faqir, *Ruzgar-i-Faqir*, Karachi, Iqbal Academy, 1952, pp. 169-70.
25. Malik, Hafiz (ed.), *Iqbal Poet Philosopher of Pakistan, op. cit.*, p. 28.
26. Iqbal Mohammed, *Six Lectures on the Reconstruction of Religious Thought in Islam*, Lahore, M. Ashraf, 1944, pp. v-vi.
27. Iqbal Mohammed, *Dar-i-Kalim*, Lahore, Ghulam Ali, 1959.
28. Shamloo (Latif Ahmed Sherwani), *Speeches and Statements of Iqbal*, Lahore, Al-Minar Academy, 1995, pp. 200-07.
29. *Ibid.*
30. Malik, Hafiz (ed.), *Iqbal Poet Philosopher of Pakistan, op. cit.*, p. 23.

2

METAPHYSICAL FOUNDATIONS OF IQBAL'S POLITICAL THOUGHT

This chapter intends to examine in a modest, critical manner some of the basic ideas which govern the political philosophy of Iqbal. A modest, critical manner because I believe a great poet philosopher, like Iqbal needs to be studied with humility yet, at the same time, with a critical attitude without which a study turns into a kind of apology, bordering on idolatrous worship. A serious thinker, like Iqbal, does not need and could never have asked posterity for such an approach.

Iqbal demands serious attention because his philosophic poetic genius represents, on the one hand, a continuity of tradition and, on the other, an attempt to integrate it with contemporary world thought with an intent set the tradition moving in an upward direction. Iqbal is unique among the contemporaries of his culture as he combines tradition with modernity with an eye on futurity, which in his peculiar symbolism is named eternity. The concept of the immanence of future, as a directing energy or a trend which operates in the manner of 'Amr' (Command), is, one of the leading ideas of Iqbal in his religio-philosophical reflection.[1] In other words, Destiny operates as an immanent factor in the world of man.

To make the present discussion precise and meaningful, it is proposed to describe the philosophical presuppositions which also operate as Iqbal's primary vision of Reality. These presuppositions are, of course, related to the spiritual culture to which he belonged, yet it will be presumptuous to believe that the entire culture consciously participates in them. It will be more precise to say that

Iqbal interpreted his culture, and its future possibilities, in the light of these presuppositions.

The Ultimate Ego

Iqbal originally started his thinking career as a pantheistic mystic.[2] Later, under, the influence of his Cambridge teachers, he became a theistic pluralist and criticized the pantheistic tendency of neutralizing the individuality of the ego and the solidity of sense data. He became hostile to the Platonic and Sufist tendency of exaltating contemplation. Thereafter, he found solace in the doctrines of the Koran, which he interpreted as a philosophy of absolute integral Ego, ever giving rise to multiplicity and finitude. He was a strong supporter of the theory of monotheism. He, however, did not subscribe to the anthropomorphic conception of the deity regarded as a stupendous patriarch, seated in some trans-terrestrial heaven. Like James Ward, Iqbal also did not put much emphasis on the cosmological, ontological and teleological arguments for the existence of God. He also interpreted the Koranic doctrine of the universe as a theory of creative evolution. Koran to him was supreme.

Iqbal adhered to the concept of an ultimate, spiritual reality and consequently believed in the actualization of a 'large purpose' in human history.[3] He categorically stated that, according to him, the Einsteinean theory dealt only with the structure of things and threw no light on the final and the ultimate nature of the entities which possessed that structure.[4] Reality, according to Iqbal, is pure duration in which there is a dynamic interpenetration of consciousness, vital force and eternal spontaneous teleology.[5] The teleological nature of reality implies that it is eternal spontaneous creativity and not a blind impetuous gigantic Elan Vita with unpredictable unforeseeable arbitrary courses of action. Reality thus may be regarded as eternal spiritual purposive creativity.[6] In *Payam-i-Mashriq* (Message of the East) Iqbal, reveals the hollowness of a materialistic *Weltanschauung*.

Iqbal was a philosopher of the cult of the Ego to some extent; like Fichte and Max Stirner, he preached the triumphant doctrine of the aggressive Ich (I).[7] Against the prevalent tenet of Islamic 'submission', Iqbal was a champion of the ego. He regarded God as the Supreme Ego. The finite egos are the differentiations of the absolute ego. The absolute ego is creative infinite spirit and spontaneous intensive energy. It is the whole of reality and its phases of life are immanently self-determined. It is rationally directed creative vitality. The attribution of egohood does not imply anthropomorphism but emphasizes that the element of life is "an organizing principle of unity; a synthesis which holds together and, focalizes the dispensing dispositions of His living organism for a constructive purpose".[8] The egos do not have their individuality extinguished in the absolute ego but receive a more pointed definition and direction. God responds to man's reflection and prayer, for "the real test of a self is whether it responds to the call of another self".[9] God is absolute because He is all inclusive and there is nothing outside Him. The philosophy of *Tauhid* as accepted in the Koran is based on the acceptance of the unity, the matchlessness and the unborn nature of God. Thus, the absolute ego is both immanent and transcendent and is a person. Iqbal wrote: "The creative energy of Ultimate Ego, in whom deed and thought are identical, function as ego unities....Every atom of Divine energy, however low in the scale of existence, is an ego. But there are degrees in the expression of egohood. Throughout the entire gamut of being runs the gradually rising note of egohood until it reaches its perfection in man. That is why the Quran declares the Ultimate Ego to be nearer to man than his own neck vein. Like pearls do we live and move and have our being in the perpetual flow of Divine Life".[10]

The ultimate absolute ego is omniscient, omnipotent, everlasting and is constantly manifesting its creative possibilities. Iqbal is very careful, however, to deny any suggestions of the least degree of pantheism that could follow as an implication from God's omniscience.

The Philosophy of Time: Iqbal and Bergson

The universe is an organic system of interconnected events, pulsating with life and energy. It is a process of continuous origination and manifestation. It is not a block structure, with its creative possibilities exhausted, and lying static in a spatio-temporal location. Iqbal formulated his metaphysical concepts at a time when Einstenian Relativity and Planck's Quantum Mechanic were dominating the world of science, Iqbal says: "........the concept of matter has received the greatest blow from the hand of Einstein...whose discoveries have laid the foundation of a far reaching revolution in the entire domain of human thought."[11] He was also partly familiar with Whiteheads theory of organism and the Russellian theories of sense-data. He also refers to Cantor's theory that space and time are continuous.[12] He was also familiar with the Bergsonian notions of creative evolution and the ideas of Bradley, Spengler, and others of their ilk. The modern researches in physics have made obsolete the old concept of matter as dense, hard substance projected in space and time. He wrote: "There is, therefore, no such thing as a self-subsistent materiality of the classical physics".[13] "It (the theory of relativity) destroys, not the objectivity of Nature, but the view of substance as simple location in space—a view which led to materialism in classical physics. Substance for modern Relativity Physics is not a persistent thing with variable states, but a system of interrelated events".[14] Iqbal maintains the objectivity of the external universe but denies its material location in space and time. He, however, departs from the trend of the mechanistic methodology of science in attributing a teleological character to the world process. He even goes a step further and like a subjectivist calls matter as 'spirit in space-time references.' His ideas bear the vital and considerable imprint of the significant trends of western science and philosophy. He accepted from modern physics and philosophy the notion of the neutralization of the substantiality of matter.

Iqbal believes in a dynamistic view of the universe. To him, nature is not a static, closed entity. In place of the notion of fixed futurity, Iqbal sponsors the concept of creativity almost in line with Whitehead. Nature is a growing process of interconnected events. There is never-ceasing growth. Matter is a colony of egos of inexplicit continual levels and out of it emerge ego of a higher order. Iqbal also accepts the Einstenian conception of a finite but boundless universe and traces this idea to the Korans view that the universe is liable to increase. The Islamic view of universal dynamism, according to him, arose as a protest against the Aristotelian view of a static universe. The Greeks were speculators and yearned after types, patterns and theories. The Islamic revolt signified a quest for the concrete, the factual and the historical. Thus, instead of the geometrical view of the fixity of the universe, there arose in the shape of the Islamic revolt a view of the universe as ever unfolding new creative possibilities. Nature, hence, is the field and the medium for self-realization. All manifestation is a revelation of the splendour of the secret self. Being an exponent of the concept of creativity, Iqbal repudiates the Nietzschean 'mystery of eternal recurrence' as mechanistic and fatalistic.

Time is real, according to Iqbal. He said that Islam sees in temporal progression a symbol of reality.[15] He repudiates McTaggart's conception of the unreality of time; he says that McTaggart's objection, that since every event in some sense is past, present and future, with reference to the perspective of the human subject who is forming these ideas, is relevant only to serial static time and not to pure duration which regards the future as an open possibility.[16] The Greek and the Hindu view of the circular movement of time is also in disfavour with Iqbal. Real time or duration is, according to Bergson, different from serialized straight linear specialized time. The ordinary conception of time or mathematical time is measured in terms of past, present and future. Iqbal accepts this Bergsonian distinction of spatialized time and duration. The real time is apprehended by the deeper inner self or what Iqbal calls the 'appreciative self'. The time of the appreciative self is single 'now'. The commonplace multiple experiences of people, on the other hand, are based on the notion of succession which is apprehended by the intellectual or empirical self which operates according to the psychological law of association. The efficient self pulverizes the single 'now' into a series of 'nows'. In mystic intuition, there is an experience of the unreality of serial time but that does not imply a complete disjunction from serial time.[17]

Although Iqbal accepts the Bergsonian distinction of serial or spatialized mathematical time and pure duration which is change without sequential succession, he differs from the French philosopher in two main respects. First, while Bergson is a vitalist, Iqbal is a spiritualist. Reality to the latter is not a vital force, but a spiritual being. It is not a blind substance but a resplendent personality and being. Second, while Bergson repudiates the teleological character of reality on the consideration that teleology makes time unreal, Iqbal imparts a purposive nature to reality. He reads the idea of a selective time process bringing out ever fresh possibilities in nature, in the simple statement of the Koran that the universe is liable to increase. He regards the absolute ego as an organic being in which there is an interconnectedness of thought, life and purpose.

The Human Ego: Freedom and Immortality

Iqbal accepts the Koranic views regarding the human ego.[18] It has a beginning in time and did not pre-exist it, except, perhaps as an ideal possibility, prior to its emergence in the spatio-temporal structure. The ego does not return to the earth after the death of the physical being. The ego is finite, but finitude is no cause for anguish. It is an opportunity for action and gradual growth in self-possession, uniqueness and intensity. Self-affirmation does not bring self into manifestation. Through others, the ego relishes, the pleasures of strife and opposition. The finite ego is not an illusion. In his poem entitled *The New Garden of Mystery,* Iqbal says:

> "The 'I' is invisible and needs no proof!
> Think awhile and see thine own secret!
> The 'I' is Truth, it is no illusion.
> When it ripens, it becomes eternal!"

The ego is in a state of constant tension to attain personality. Through hard struggles, it gains the capacity to experience freedom. Its real nature is that it is not only the permanent subject of psychic experiences, but is a spontaneous creative onward movement. In the *Asrar-i Khudi,* Iqbal says:

> "It (i.e. self) is staying by the strength of its arm
> That it may become conscious of its own strength.

> Like the rose, it lives by bathing itself in blood.
> For the sake of a single rose it destroys a hundred, rose-gardens .....
> The excuse, for this wastefulness and, cruelty
> Is the shaping and perfecting of spiritual beauty ....."[19]

According to the *Javed Nama* of Iqbal, there are three steps in the evolution of the ego: (i) The realization of the creative possibilities of the ego is the first stage. It is the stage of individuality, (ii) The second stage is to see the ego in the context of other egos. This may be called the stage of sociality, (iii) The third stage is to realize God and to view the ego in the perspective of this divine consciousness. But being steadfast before the eternal light of God, one can become powerful and, thus partly feel like the omnipotent God.

The ego is not immortal but has the capacity for immortality. Immortality is a persistent aspiration and a constant ideal for the ego, rather than a realized actuality. It is a goal and not a possession. If man does not make efforts, he can sink down to the level of dead matter. Through persistent endeavours alone in maintaining the tension of life, the ego becomes a candidate for immortality. Iqbal says: "When the 'I' ripens into a self it has no danger of dissolution."

Religious Experience

It is possible to attain the experience of the spiritual real through intuition. Intuition, however, is not super-rational but is of the type of cognition and volition. In the very act of knowledge, thought transcends its finitude of definite specifications and attains an immanent infinitude in its dynamic self-expression. There is no opposition between thought and intuition. The intuition of the self reveals its freedom and immortality. From the intuition of the self which is basic concrete experience and which reveals life as a centralizing ego, it is possible to advance in spiritual experience and obtain the intuition of the absolute reality. Intuition is direct, immediate and integral experience of God as concrete being. In mystic experience, the moral distinctions between the subject and the object cease. It is a state of deep association with a unique other and higher self. Due to its unanalysable wholeness, the intuitional mystic experience is incommunicable in predicative terms, but its incommunicability is no ground for its non-objectivity. In different ages and countries, a number of persons have testified to the reality of their experience. It is the faculty of heart or what the Koran calls *Faud* or *Qalb*. Intuition and rational thought complement each other because intuition is the deepening and elevation of thought.[20] If intuition reveals the eternal nature of reality, rational thought seeks to probe into the phenomenal sphere with its series of definite specifications and diverse frames of reference.

Prayer: Its Social and Political Implication

Cosmopolitanism is the necessary political implication from the concept of a supreme ego. Thus, "from the unity of all inclusive ego, who creates and sustains all egos follows the essential unity of mankind".[21] According to Iqbal, the differentiations of mankind into tribes, races and nations are only for the pragmatic purpose of identification. There are no intrinsic and inherent rankings and gradations. Prayer has a moral purpose. It expresses the yearning of man for association with the supreme ego. Reality responds to man in the act of worship. But collective prayer has a wider sociological aspect too. It expresses social solidarity. The daily act of prayer asserts the social unity of man. The annual ceremony at Mecca expresses Islamic fraternity in concrete shape. Thus, prayer has a moral and a political aspect as well.

The Spirit of Muslim Culture

Iqbal repudiates the concept of pure contemplation and the attitude of ignoring the factual and the concrete. He condemned the mystic minimization of the role of intellect as a factor responsible for the decadence of the oriental nations. He pronounced hostile judgment on the role of Greek speculative influences on Islamic thought. Hence he wanted to revive the vigorous spirit of early Islam. He feels that the ascetic other worldliness and, the repressive devotion to Sufism generated an attitude which missed the Islamic emphasis on the realization of the spiritual through the temporal. In the *Asrar-i Khudi*, Iqbal criticized the teachings of Hafiz Shirazi because he taught inaction. The negation of the temporal leads to the social stagnation. The distinction between *Zahir* and *Batin* (appearance and reality), as propounded in some aspects of Sufist thought, inculcates an attitude of indifference towards the world and its pressing problems. He, on the other hand, advocates constant participation in group and communal activities. Religion is there to integrate the forces of man's life and to make him a creative participant in the drama of history by revealing to him the unplumbed depths of inner existence.

Iqbal appreciated the doctrine of *Karmayoga* taught by Sri Krishna but he was hostile to the dialectics of Samkara. He wants that writers and social leaders should lead a life of activity, liveliness and quickness. He says:

> "If thou hast the coin of poesy in thy purse,
> Rub it on the touchstone of life;
> For, along time thou hast turned about on a bed of silk,
> Now accustom thyself to rough cotton!
> Now throw thyself on the burning sand,
> And plunge into the fountain of Zemzem.
> How long wilt thou fain lament like the nightingale?
> How long make thine abode in gardens?
> O thou whose auspicious snare would do honour to the Phoenix,
> Build a nest on the high mountains
> That thou mayst be fit for life's battle,
> That thy body and soul may burn in life's fire!"[22]

The gospel of Islam as interpreted by Iqbal is one of activity and strength. He says that the teaching of the Koran is meliorism based on the recognition of a growing universe. Action contains the pith of life. The world is not a perpetual vale of tears, nor is it a crown of roses. The Islamic teaching as interpreted by Iqbal, is opposed to the mechanism and rigidity involved in Nietzsche's concept of 'eternal recurrence'. From the concept of the ego there follows the notion of a sturdy individualism in Iqbal. He would not put much reliance on the sheer force of numbers. The sole sources of strength in society are self-reliant individuals or what Iqbal calls 'self-concentrated individuals.'[23] They alone reveal the immensities and profundities of life. They have an intuitive grasp of the fundamental bases of existence and they perceive the new criteria of social evolution and adjustment.

The Superman

Like Nietzsche, Iqbal is an exponent of the concept of the superman. His superman (*Insan-i-Kamil*), however, is not a 'titanic Nordic blonde beast' representing the aristocratic virtues of Prussian Junkerdom. Supermanhood is dependent on a process of self-discipline. Conceit and self will have to be renounced. He who can control himself can wage a successful fight with the elements and can wear the 'Crown of Solomon'. The forces of the world will have to yield supremacy to the man of strong iron determination. This superman as conceived by Iqbal delights in being the vicegerent of God (*Niyabat-i-Ilahi*). He is an effective instrument for the execution of the will of God. Thus the two characteristics of the Iqbalian superman are: (a) absolute self-control, and (b) willing obedience to the ordinances of God. Iqbal says:

> "God's vicegerent is as the soul of the universe,
> His being it the shadow of the greatest name."

The important presuppositions of Iqbal's poetic philosophy, which, in a direct and sometimes in an indirect manner, govern his political thinking, are: (i) reality is spiritual and its life consists in its temporal activity, on account of which all that is secular is sacred in the roots of its being; (ii) the world of nature, i.e. the spatio-temporal order, and the world of man are continuous, the middle term being self or ego; (iii) finite human ego is so far the highest instance of cosmic evolution, and that it is free; and (iv) the direction of human history is the transformation of man into a more perfect being, the vicegerent of God, being the actualisation of the original Trust. The last item transforms anthropology into theodophy, the ultimate aim or intention' of history being identified with the emergence of 'gaostic beings. It also gives to history an eschatological dimension. In the words of Alam Khundmiri an eminent scholar, "I call them presuppositions because Iqbal never worked them out in a rigorous philosophical manner, neither gave reasons, if reason is to be distinguished from analogy, as to why they should be regarded as true. They can more appropriately be regarded as mystico-poetic visions, or as elements of his utopic consciousness, immanent in his poetic creations and philosophic reflections. I have used the word utopic deliberately to distinguish it from utopia, an ideal construct or an ideal type. It is the presence of these elements that gives a mystic dimension even to Iqbal's political philosophy and eventually fails to relate it to the concrete problems of his contemporary age."[24]

In the light of these presuppositions, it is proposed to discuss the following concepts: Iqbal's idea of the transformation of the human world, the agents of this transformation or the problem of leadership, and the meaning of freedom in the world of man. The first had always been one of the dominant concerns of Eastern spirituality to which the problem of human conduct had been related and which gives meaning and significance to political action. To an 'outsider', it might seem preposterous that they are termed as problems of political philosophy, but they acquire significance in the Eastern tradition which gives primary importance to spiritual salvation. Since Iqbal belongs more to the mystical tradition of Islam than to its politico legal one, this observation acquires significance.

The Islamic mystic tradition, like all such traditions, gives priority to the vertical dimension, of human personality, at the expense of his horizontal dimension, comprising the social and political conditions of his existence and leaving them to historical contingency.

Iqbal's significance as a thinker lies in the fact that he realised the importance of the social and political conditions of existence in the scheme of spiritual perfection; recognised the existential tension and made a sincere, though unsuccessful, attempt to reconcile the two as complementary poles of human existence. Islamic religious tradition which, after the convergence of the Asharite theology and mystic illumination in the personality of Al-Ghazali, had become predominantly the centric, was given an anthropocentric shift by Iqbal, by declaring that Man is a co-worker with God. Since religious consciousness repels complete anthropocentrism, it leads to a tension between the two in Iqbal's religio-political consciousness, a tension which also becomes a source of tragic feeling in the last phase of Iqbal's poetry.

The Transformation of the Human World

Iqbal's attitude to this problem has to be viewed from two levels which also involves a certain tension between them. One is the fact of change and which for Iqbal, in the tradition of the contemporary biological thought, has the status of a natural law governing the entire universe. It is the fundamental concept which makes the idea of transformation possible. It is to be noted here that Iqbal, like Kierkegaard, accepts change as a fact of the human world having a religious and moral perspective before him. It was not the problem of change as it occurs or becomes as such with which he was concerned, but rather the problem of change as it occurs to one who is involved in becoming oneself and this process of becoming oneself, in turn, involves a distinction between what one is and what one ought to be. In other words, transformation involves the idea of 'not yet' which is one of the recurring themes in Iqbal's poetry. The entire universe along with the world of man awaits, at every moment of its life process, a stage which it has not yet arrived at. Iqbal sees even in twentieth century science and particularly in the Einsteinian theory of the 'expanding universe' a powerful argument to support his view of perpetual growth, an instance which goes to show that he was not arguing in a rigorous philosophical manner but seeking analogies to make his point of view tenable. The idea of growth and expansion when combined with his presupposition, that the universe is spiritual, leads him to believe that it is becoming more and more spiritual or acquiring greater spiritual dimensions. It makes the transition to the ideal of human transformation not only possible but in a certain sense obligatory for man. Man must develop his personality for his ego by transcending the finite material conditions and the spatio-temporal relations, the essential components of his existence.[25] He believes, like Bergson that matter is a necessary hindrance to be overcome in the process of evolution and that it does not have an independent ontological status. Matter itself being a lower type of ego activity, it offers resistance to the conscious activity of the higher ego which aspires for progressive expansion of consciousness leading to personal immortality. In one of his significant notes in *Stray Reflections*, he declared that personality, being the dearest possession of man, must be looked upon as the ultimate good which must work as a standard to test the worth of our actions. Needless to say that he was influenced by Nietzsche but, again, a case to prove that he had motives different from those of big 'masters'. The existence of the vital impulse makes spiritual transformation not only possible but also serves as a necessary condition for this activity. It solves the riddle why Iqbal believes in the primacy of will over intellect; intellect or reason gives meaning to or seeks meaning in the past; it is will which does so to the future. Whereas the non-existing past acquires meaning through reason, the non-existing 'not yet' gets it through will. The flux of time, an important notion of Iqbal, does nothing by itself. It is only through the mediating agency of will that transformation becomes possible. Reality of time serves as an essential precondition to make spiritual growth possible. In other words, the universe must be constituted in such a manner that time, however logical, is its necessary component. Like Kierkegaard, Iqbal believes that existence, i.e. the mode of finite existence, is future oriented, and the hallmark of existence is the forward movement. If the Socratic maxim was 'know thyself' from within, the Iqbalian maxim is 'change thyself from within'. It is only in the process of changing oneself that one gets to know oneself: cognition follows the act of will. It has to be noted here that the nineteenth century offers two perspectives of human transformation: one is the Marxian perspective, shared by the later sociologists that man changes himself in the process of changing the world and also gets to know himself in the process of change. It is what Marx says, existence precedes consciousness. The other perspective, that of the existentialists and the vitalist religious thinkers, and shared by Iqbal, is the precedence of inner change over external institutional change.

Iqbal calls the former, borrowing a term from Spengler, the intellectual way of making the world our own which consists in understanding the world in a causal manner, and the latter as the vital way of appropriating the universe, which according to him, is what the *Koran* calls *Iman*. This vital act too implies a necessity, but different from it. It is what Iqbal calls 'higher fatalism', without making clear what is meant by it.

The idea of *Iman* leads to the important notion that transformation is the acquisition of *Iman*. This vital act results in the complete appropriation of the cosmos or in the personality of the agent (*Momin*), the reverse of which is the absorption of the human personality in the material universe; in other words, the personality loses itself in the spatio-temporal order called serial time.

The phraseology reminds one of the authentic and inauthentic of Heidegger, the corresponding states being existentiality and facticity or fallenness.

It is important to note here that in our passion to make Iqbal 'contemporary', we may ignore the danger of giving these phrases a scientific orientation. This vital act is also called rebirth in the *Javed Nama.*

This symbolism has a close relation to the prophetic symbolism of the splitting of the chest; an event which preceded the Ascension. Corresponding to this transformation on the level of the individual personality is the ideal of the transformation of the human collectivity. This process of the transformation of humanity has to be preceded by what Iqbal calls 'Recons Resurrection'.

> "He from whosc body the pure spirit has departed
> Cannot rise from the dust without a Reconstruction".[26]

The word 'revolution' which occurs often in Iqbal's poetry has a certain resemblance to Marxian terminology, but it has also to be noted that Marx had secularised the Semitic archetypal symbolism of Reccons Resurrection in his revolutionary vision. This resemblance, although a deeper one, has not to bc stretched too far. In the political philosophy of Marx, the process 'transformation' follows revolution, the destruction of the old forces of production and relations, whereas for Iqbal transformation alone guarantees a new social order, as it has once happened during the Prophetic mission.

This notion of transformation leads to another significant symbol of the 'new man', a passion common to the vitalistic philosophies of Nietzsche, Kierkegaard and also of Marx. Iqbal blends his vision of the 'new man' with the theosophic symbol of *al-Inan-al-Kamil*, with one important difference that for the Islamic theosophists, perfect man represents the essence of creation, logically preceding the creaturely order, whereas for Iqbal it seems to lie in future, so far as its collective manifestation is concerned, although its individual prototype lies in the personality of Mohammed. Evolutionary theory in this case, too, serves Iqbal's mystical and gnostic purposes. A corresponding idea is found in the current Christian theology, the passion to create the Kingdom of God here and now, by intervening in the process of history.[27]

Is it a mere fortuitous event that while this notion looks exhilarating on the plane of the individual personality, it leads to depressing consequences on the socio-collective level, since it promotes a rather backward looking revivalistic attitude? It is neither accidental nor a mere wrong interpretation of Iqbal by his revivalist following. This trend was, on the contrary, implied in his vision of the future, because the 'utopic'.elements of his vision have a direct reference to the 'Golden Past' wherein lies the lost utopia, and it is this reference which makes his symbol of resurrection dangerously meaningful. One radical element of his vision was, however, lost sight of by those who sought practical guidance from Iqbal, in spite of the Quranic injunction not to follow the poets, and that was his notion of continuous growths even after Resurrection. If Resurrection leads to growth on the individual psychic plane, it follows that the same ought to happen on the plane of collective life. This means that reference to past must lead to an evolution in future, which in turn implies that the real 'historical ideal' lies in the future and the point is that for Iqbal, the future is not unrelated to eternity, the plane of unrealised possibilities, but a little reflection shows that this analogy is at best a poetic metaphor, since the spatio-temporal world, the stage of history, is different in its nature, from a possible world, the stage of spiritual growth after 'resurrection'.

One of the significant consequences of this notion of transformation, bearing a relevance to the contemporary human situation, consists in the approximation of the secular and the religious, as maintained by Iqbal, the difference between the two does not lie in the act itself but in the attitude of mind with which the agent does it. "It is the invisible mental background," according to Iqbal, which ultimately determines its character. An act is temporal or profane if it is done in a spirit of detachment from the infinite complexities of life behind it: it is spiritual if it is inspired by that complexity".[28] This approximation works effectively only on the ideal plane. The facts of political life are, however, such that unless the political act is performed by one who is completely transformed, a prophet for instance, the balance cannot be maintained. Since such moments are very rare in human history, the profane, as it tends to relax the tension of the agent, assumes the guise of the religious or the spiritual, the result being the psychoanalytic procedure of rationalization, which can be called the greatest danger to human civilization and also to spiritual life.

Agents of Transformation

This observation leads us to the second problem of the agents of transformation or of leadership, one of the most crucial aspects of Iqbal's politico philosophical vision. Since the later developments of a philosophic genius lie in his primary vision of reality, the question of Iqbal's early vision becomes significant. There is a strong evidence to prove that Iqbal 'perceived' Reality in its aspect of Power, and the tremendum overshadowed the fascinosum. The poetic creations belonging to the comparatively mature period also provide a testimony to this assumption. As early as 1910, he commented: "Power is more divine than truth. God is power, be ye then like your Father who is in heaven." He goes to the extent of declaring, "Power toucheth falsehood, and lo, it is transformed into truth."[29] Nietzsche was, of course, a powerful influence, but the important psycho-biographical point is that Iqbal's perception was also responsible for drawing him towards Nietzsche. Power is one of the most recurring themes in the poetry of Iqbal.

The idea of a powerful personality may also be considered an archetypal symbol which arouses conscious admiration at certain critical moments of history, and the nineteenth century seems to have provided a favourable cultural soil to this idea. It was not only Nietzsche, it was Marx, too, who accepted this notion and transformed it into a class; it is also implicit in the religious philosophy of Kierkegaard in the form of the authentic person who can take a decisive leap. Rousseau, with his idea of the noble savage, seems to have initiated this powerful romantic philosophic movement and it is on this account that some sort of 'neo primitivism' in some form or other persisted as an essential component of the contemporary radical political philosophy. In India, too, the literature of the nineteenth century, in Bengal and Maharashtra, eulogised power, which inspired militant nationalism.

Iqbal, too, believed that a great personality alone could revitalize a dying social organism by the revelation of a new ideal. While Marx saw such new ideals imminent in the current historical situation which can be grasped in the process of social action, to Nietzsche the discernment of these ideals is the result of an inner vital contemplative act possessing an annihilating quality. Iqbal shares this belief with Nietzsche, with the difference that the discovery of new ideals is an outcome of a mystical experience when it "tends to overflow its boundaries, and seeks opportunities of redirecting or refashioning the forces of collective life". In this experience, according to him, "the finite centre of life sinks into his own infinite depths only to spring up again, with fresh vigour to destroy the old, and to disclose the new directions of life." The reference here is to the Prophet's experience, but it is to be noted that the Prophet, for Iqbal, is more a model for humanity than one who merely belonged to one particular moment of past history of mankind. In Bergsonian language, the higher ideal must incarnate in a privileged person who becomes an example. Iqbal had a glimpse of the later Bergsonian idea in his earlier poetic symbol of the 'free man' who has to be a model for the 'Bondsman', an idea which was implicit in Bergson's Time and Free Will.[30]

It is the 'free man' alone who can initiate a new social order and restructure the existing institutions, Iqbal believed. He has a strong belief that self-concentrated individuals are the only effective power to counteract the forces of decay and such individuals alone reveal the depths of life. It is only through them that new standards are disclosed in the life of which we begin to see that our environment is not wholly inviolable and requires revision. It is upon such self-concentrated individuals that the title of 'Man of Faith' is conferred. He is also called *Qualandar*, the archetypal symbol, whose important attribute is voluntary poverty. It is the privilege of such persons to enjoy 'pure duration' through their intuitive power and have a glimpse of the Eternal. The inference is clear that only those, who have 'discovered' the nature of Reality, can bear the burden of changing the world. They alone are those who carry the Trust.

This notion too works quite neatly on the ideal plane but the question remains unanswered as to how they can be induced to act as agents of change. Iqbal faces the same difficulty which was once faced by Plato and since Iqbal's passion to transform the existing reality is so strong and his faith in powerful persons so unshakeable, that he had to create certain myths. The unfulfilled mission of Islam was one such myth and to perceive it as real, he endows this sort of spiritual greatness on such historical persons as Nadir Shah and Abdali, because they unified the Iranians and the Afghans. He could tell his western counterparts that we too had our Bismarcks and Mazzinis. In his heavenly journey, Abdali is recognised as:

> "The leader of all the martyrs of love
> Glory of India, China, Turkey and Syria
> Whose name is more resplendent than the sun and the moon,
> The dust of whose grave is more living than I and you".[31]

A serious question arises here, as to how one can judge the authenticity of such great men who are supposed to be the guardians of human destiny. Iqbal offers the pragmatic test. He can believe that it works satisfactorily when it is applied to a Prophetic personality.

But can it work as effectively in relation to Nadir Shahs and Abdalis? It is a question which Iqbal never asked or answered. History is a poor judge to tell us who was and who was not a false Prophet, since historical judgements are never completely free from subjective preferences and secondly all pragmatic tests have a reference to historical time which makes it difficult for a judge to look at this question from the perspective of eternity, as eternity alone changes history into simultaneity and, therefore, historical judgements involving a reference to value remain ambiguous.

Meaning of Freedom

Coming to the problem of freedom, one may feel tempted to ask if freedom and power, as embodied in a powerful personality like Nadir Shah, or Abdali can go together. The answer is both yes and no. If the word freedom is used in the sense in which it is used in ordinary language, the answer will be a simple no, because the freedom of the powerful person puts limits on and tends to negate the freedom of the rest. The answer is yes, since power cannot be achieved without the possibility of freedom in the universe. A little reflection reveals that the situation, however, is not so simple, and since the terms, power and freedom, carry some ambiguity, the intention of the speaker has got to be studied. It becomes much more imperative when the speaker happens to be a poet also.

Undoubtedly, Iqbal had a great passion for both power and freedom. His selection of characters in this context in *Javed Nama* indicates that there was a shift in the meaning of the word 'Sultan'. In the beginning of the poem, he meant spiritual power and at the penultimate stage absolute political power was meant. The other word used by him for power so often in his poetry is *Qushiri*, as opposed and complementary to *Dilbari*. The term *Khudi* itself carries both the qualities of power and of freedom, and hence it has not to be confused with the term *Atman* or simple *Nafs*. The simple subject 'I' acquires egohood when it is free and powerful, and acquires them through Love or *Ishq*, which is a vital force. Since the vital force is the mediating term, good becomes identical with the vital force and consequently freedom loses its ethical content and becomes synonymous with freedom to acquire power which means it cannot be shared. Bergson and Iqbal present an interesting parallel as Bergson's open morality transcends precepts, depends upon models, on lives which are sources of inspiration, on experiences not amenable to discursive expression, so is Iqbal's notion of freedom. It can be 'enjoyed' by a powerful Ego, but cannot be shared among unequal 'citizens'. It is on account of this combination of power and freedom that Iqbal has reservations about democracy, although he charges Nietzsche with advocating an aristocratic model. Iqbal is sincere in his charge against Nietzsche as he has a different model in mind and that is the Prophetic model; nonetheless the relationship between a prophet and the people can only be of command and obedience; hence Iqbal's insistence on obedience having a moral value.

Had it been a stray instance, it could have been ignored, but since it is one of the recurrent themes, it has to be regarded as a dominant passion of Iqbal. Passion for power creates the 'pathos of obedience', an ecstatic joy in surrendering one's own freedom without coercion. This ecstatic joy produces the 'politics of ecstasy' if instead of enjoying it in an inward manner, which is the true function of poetry, it becomes a guide for action also. One speaks of Iqbal and Sri Aurobindo as cures for the contemporary dehumanisation caused by the scientific culture, but it is not to be forgotten that this anxiety against this phenomenon of dehumanised culture might produce some sort of neo-primitivism, which instead of humanising our behaviour makes it more irrational, hence ecstatic, giving rise to a 'culture of ecstasy'. The danger is there. It vindicates the Quranic maxim that poets are not to be followed, they might take us to hell. Not only good intentions, sweet voices also pave the way to hell. Such voices are to be enjoyed, contemplated upon, but are not to be taken too seriously, if they go against the sweeter common sense.

REFERENCES

1. S. Alam Khundmiri, "A Study of the Concepts of Transformation, Leadership and Freedom in the Political Philosophy of Iqbal" in Jafri Ali Sardar & K.S. Duggal (eds.), *Iqbal Commemorative Volume*, New Delhi, All India Iqbal Centenary Celebration Committee, 1980, p. 93.
2. Iqbal Mohammed, *Six Lectures on the Reconstruction of Religious Thought in Islam*, Lahore, M. Ashraf, 1944, pp. 76.
3. *Ibid.*, p. 233.
4. *Ibid.*, p. 52.
5. *Ibid.*, p. 75.
6. *Ibid.*, pp. 70-72.
7. Varma, V. P., *Modern Indian Political Thinker*, Agra, Laxmi Narayan Aggarwal, 1961, p. 441.
8. Iqbal Mohammed, *Six Lectures on the Reconstruction of Religious Thought in Islam, op. cit.*, pp. 82.
9. Dar, B.A., *A Study in Iqbal Philosophy*, Lahore, M. Ashraf, 1944, pp. 402-13.
10. Iqbal Mohammed, *Six Lectures on the Reconstruction of Religious Thought in Islam, op. cit.*, pp. 99-100.
11. *Ibid.*, p. 47.
12. *Ibid.*, p. 50.
13. *Ibid.*, p. 52.
14. *Ibid.*
15. *Ibid.*, p. 54.
16. *Ibid.*, pp. 77-78.
17. *Ibid.*, p. 29.
18. Dar, B.A., *A Study in Iqbal Philosophy, op. cit.*, pp. 396-401.
19. Iqbal Mohammad *Asrar-i-Khudi*, Lahore, Ghulam Ali, 1957 (Translated) as '*Secrets of Self*', New Delhi, Arnold-Heinemann, 1978, p. 35.
20. Iqbal Mohammed, *Six Lectures on the Reconstruction of Religious Thought in Islam, op. cit.*, pp. 23-31.
21. *Ibid.*, p. 129.
22. Iqbal Mohammad *Asrar-i-Khudi,* Lahore, Ghulam Ali, 1957 (Translated) as '*Secrets of Self*', New Delhi, Arnold-Heinemann, 1978, pp. 28-29.
23. Iqbal Mohammed, *Six Lectures on the Reconstruction of Religious Thought in Islam, op. cit.*, p. 212.
24. S. Alam Khundmiri, "A Study of the Concepts of Transformation, Leadership and Freedom in the Political Philosophy of Iqbal" in Jafri Ali Sardar & K.S. Duggal (eds.), *Iqbal Commemorative Volume, op. cit.*, p. 94.
25. Iqbal Mohammed, *Six Lectures on the Reconstruction of Religious Thought in Islam, op. cit.*, p. 19.
26. S. Alam Khundmiri, "A Study of the Concepts of Transformation, Leadership and Freedom in the Political Philosophy of Iqbal" in Jafri Ali Sardar & K.S. Duggal (eds.), *Iqbal Commemorative Volume, op. cit.*, p. 97.

27. Malik, Hafiz (ed.), *Iqbal Poet Philosopher of Pakistan, op. cit.*, pp. 204-205.
28. Iqbal Mohammed, *Six Lectures on the Reconstruction of Religious Thought in Islam, op. cit.*
29. S. Alam Khundmiri, "A Study of the Concepts of Transformation, leadership and Freedom in the Political Philosophy of Iqbal" in Jafri Ali Sardar & K.S. Duggal (eds.), *Iqbal Commemorative Volume, op. cit.*, p. 98.
30. Malik, Hafiz (ed.), *Iqbal Poet Philosopher of Pakistan, op. cit.*, pp. 202-203.
31. S. Alam Khundmiri, "A Study of the Concepts of Transformation, Leadership and Freedom in the Political Philosophy of Iqbal" in Jafri Ali Sardar & K.S. Duggal (eds.), *Iqbal Commemorative Volume, op. cit.*, p. 100.
32. *Ibid.*, p. 101.

3

DEVELOPMENT OF POLITICAL PHILOSOPHY

Iqbal's eminence as a poet has been acknowledged; however, his status as a political philosopher is a controversial matter. An important reason for this controversy is that Iqbal's thought has been studied in parts but not as a whole. In order to appreciate a work, whether political or philosophical, it is necessary to analyze the elements that compose the whole; moreover, one must consider the relation of the parts to each other and to the whole. Iqbal's ideas have been divided into fragments but the elements have not been synthesized.

This has resulted in the general impression that Iqbal had certain ideas of philosophical significance, but that these ideas do not form a system. It is important to understand the underlying pattern of Iqbal's thinking and what he envisioned for only then is one able to understand the true significance of his ideas.

Concerning his political philosophy, Iqbal is often accused of inconsistencies and contradictions. He has been misunderstood. This accusation is largely the result either of regarding Iqbal's general philosophy as arising from his political philosophy or of studying his political philosophy in isolation from his general philosophy. Actually, Iqbal's political philosophy is an integral part of his general philosophy and is best understood if studied with reference to the wider whole.

What is known as Iqbal's political philosophy is usually based on his writings from 1908 to 1938. However, in order to understand the continuity of his thinking, it is necessary to pay considerable attention to the two earlier phases of Iqbal's poetic career, that is, from 1895 to 1905 and from 1905 to 1908. Most of the poems written during these two periods are found in Parts I and II of *Bang-i-Dara* [The Call of the Highway].

Pre-1905 Political Poetry: The Temple of Love

Iqbal's pre-1905 poetry is often described as nationalistic. It is necessary, however, to differentiate between patriotic and nationalistic verses because the latter implies an awareness of, and an involvement with, political theory or practice which may be entirely absent from the former. "Patriotism is a perfectly natural virtue and has a place in the moral life of man,"[1] wrote Iqbal, and he himself never lacked this virtue. *Himala*, the first poem in *Bang-i-Dara* and a hymn of the magnificence and grandeur of the tallest mountain range in the world, is patriotic: it was inspired by the beauty of the land of his birth. Like *Himala* (The Home of Snow), Iqbal's other "patriotic" poems are buoyant and heart-lifting. The fire that raged within him left on his poetry the stamp of Miltonic restlessness. There are, amongst the vast bulk of Iqbal's writings, only a few poems which impart a feeling of peace and joy. These poems reflect those moments when the poet sought an escape from the ravages of his soul in thoughts of love and sensuous beauty. That Iqbal the poet was a patriot there is no doubt. That he remained a patriot is sometimes forgotten. H.L. Chopra, a widely known scholar has truly said: "Iqbal sprang up in the realm of Indian poetry as a bard of India and even after so many vicissitudes in his career when he left this earth on 21st April, 1938 he remained a truly patriotic poet of India."[2] But it must be kept in mind that patriotism is not the same thing as nationalism.

The two things which stand foremost in Iqbal's pre-1905 political poetry are: (i) his desire to see a self-governing and united India—a country free of both alien domination and inner dissension, in particular the Hindu-Muslim conflict; and (ii) his constant endeavour to draw attention to those factors of decadence which caused the decline of the Muslims in India.

The political poems in the first part of *Bang-i-Dara* indicate that Iqbal suffered more on account of the factions which divided the Indian people than the imperialistic yoke imposed on his countrymen. One reason for this was, no doubt, the feeling that until the people of India could resolve their differences and come to terms with each other, the deal of sovereign rule for India would remain a dream. But there was another reason—one far more fundamental for Iqbal's passionate nationalism, which found its expression in poetic outbursts that resounded throughout India.

Iqbal the young poet was deeply disturbed by the discord and distrust between Hindus and Muslims which he saw all around him. With his usual eloquence, he cried out against the forces of disruption. No one was spared; the *Mulla* and the *Brahman* were criticized equally. He maintained that from their gods they learned only hatred and hostility ("You think in idols carved of stone there is divinity") and professed allegiance to a faith other than theirs ("Each particle of my country's dust is as a god to me").

The poet adored his country, but its places of worship were defiled by hatred; hence he believed a new temple, a sanctuary of love, had to be built wherein all could worship. In *Naya Shiwala* (New Temple) the poet addresses the keepers of the temple and the mosque:

> Come, let us lift the curtain of Apartness once again,
> Unite once more the parted ones, erase duality's stain.
> Since a long time heart's habitat has been so desolate,
> Come, a temple new, in this land, let us elevate.[3]

Although the theme of this poem recurs in Iqbal's early verses, it is unique because it gives a message of universal love not in general terms as it appears elsewhere in Iqbal's poetry, but in specifically Hindu vocabulary. This poem was a direct appeal to the Hindus, and its terminology indicates the poet's sincere desire to establish intellectual communication and spiritual rapport between Hindus and Muslims.

In *Taswir-i-Dard* (Portrait of Anguish), the poet cries for Hindustan whose people are lost "in tales of ancient days" and do not indulge even in "the joys of complaint." The poet expresses his own determination "to string together the separated beads of the rosary," the religious association of the image making it more effective in the name of an all embracing principle of human love, the poet urges his compatriots to transcend cultural and religious boundaries, attaining true freedom:

> In Love is hidden Liberty, if only you could see
> And bondage is discrimination between you and me.[4]

Iqbal did not minimize the difficulty of eliminating bias. He admired natural phenomena for their "impartiality"; nature's beauty entertained one and all. The poet was conscious, however, of his own limitations. To the *shama* (the candle), he said wistfully:

> In the Ḳa'bah and the house of idols, your glow is the same,
> But in differences of mosque and temple lost I am.

And to his compatriots he said unequivocally:

> People of India, if you do not learn, you'll be no more.[5]

The message of the poet for the people of India was that of love and faith. Many years later, when discussing the problems of India, Iqbal wrote: "It is, and has always been, a question of faith. Our faith too, depends on affection and understanding. What we need for a swift solution of the political problem of India is faith."[6] This message of love and faith must be kept in mind because Iqbal's political philosophy is an integral part of his total commitment, and to regard his words as possessing merely political significance is to elude the core of his thought.

A number of poems for children were obviously written with India's political situation in mind. *Parinde ke Faryiid* (The Bird's Lament), the longing of a caged bird for the freedom of the skies and its nest, bears reference to the greater bondage and deeper yearnings for liberty of a whole subcontinent. The vision of *Hindustani Bachon Ka Qawmi Git* (National Anthem of Indian Children) is again one of universal love, peace, and goodwill. The poet wished to create in little children the feeling that to live in India was to lead "a heavenly life", and that they must "adorn my country as a flower adorns a garden". Iqbal's *Tarana-i-Hindi* (The Song of India) became the unofficial national anthem of India.[7] According to one Indian writer, it "remains to this day the best patriotic poem written by any Indian poet in modern times. It comes nearest, in fact, to a truly non-communal national anthem of India."[8] The message it contained said nothing new, but perhaps the words were more explicit than they had been before:

Religion does not teach us mutual belligerency,
We are the people of India, and India is our country.

Perhaps it was the tragedy of Iqbal's youth that the India mother of all myth exploded before the children who sang his songs and grew to manhood during his life time. Disillusionment came, as it must when the idealism of youth is touched by the coldness of reality. But, as Keats had clung to the principle of beauty even as sickness and sorrow consumed his mortal self, although the tune of his melody became more solemn as the ease of the first fresh notes of his singing became burdened with experience.

It is difficult to say that before 1905 Iqbal had a definite philosophy. But an analysis of his early poems reveals in embryo most of the ideas which later played a prominent part in the system of his thought. For instance, one finds the idea that the development of the individual depends on his capacity for purposeful action. The destiny of a nation too depends on action:

'Tis Divinity's decree and Creation's mode —
Who walks along the path of Action is by Nature loved.[9]

"Action is the highest form of contemplation",[10] Iqbal wrote in 1925, but he knew this truth a long time before. He knew it, in fact, before the active West had sharpened his awareness of the consequences of unpreparedness and unwillingness or inability to act when action was required. Perhaps he learned to believe in action from the Koran, which "emphasizes 'deed' rather than 'idea' ".[11] But it is more likely that his love for action sprang from his particular disposition and that it was strengthened rather than induced by his religious faith.

Another interesting idea in Iqbal's early verse is that religion does not demand that a man neglect his material needs. In *Din-o-Dunya* (Religion and World), Iqbal states that in fact "religion is the guardian of the material world." In another poem, the spirit of Sir Sayyid Ahmad Khan tells the poet:

> To teach religion in the world, if this be your aim,
> Do not teach your nation that the world they should disclaim.[12]

This is the foreshadowing of the idea which later became one of the pivots of his political philosophy that the spiritual and temporal aspects of human life, far from being mutually exclusive, were organically related to each other, and that there was no need to bifurcate the unity of individual and collective life into the domains of church and state.

The Temple of Love may be said to symbolize the first phase of Iqbal's poetic career. E.M. Forster has misunderstood the vision enshrined in this symbol. He described the new temple as "the Temple of India" and observed: "The glory of the courtyard from Mecca shall inhabit that temple; the image on its shrine shall be gold, inscribed Hindustan, shall wear both the Brahman thread and the Muslim rosary, and the *Muezzin* shall call worshippers to prayer upon a horn."[13] The new temple is in India, but it is not of India. Nor is it conceived by Iqbal as heralding the advent of a new faith containing elements of Hinduism and Islam. Such a faith could not create an organic unity and might very well have destroyed the distinct identity of either faith. The temple is raised to love, which transcends the images in which it is expressed.

Also, it is asserted that Iqbal's thinking was completely changed by his contact with Western civilization. This is hardly true; Europe crystallized rather than fundamentally altered his thinking. Coming face to face with a society so different from his own, he had the opportunity to compare the modes of living and the systems of values he saw in the East and the West. No longer was it sufficient for him to believe instinctively or vaguely in the reality or validity of a certain world order. He had to develop a rational basis for his beliefs. The spontaneous outpourings of his verse had to give way to poetry based on considered judgments; in his future work he was to guard against what T. S. Eliot has called the "dissociation of sensibility."[14]

Iqbal, who had sung so many hymns to love, felt nothing more keenly in the West than its lack of love. The heart (regarded by the mystic tradition to be the repository of love and the source of true knowledge) had been displaced in the West by the mind. Iqbal was not anti-intellectual; for him, thought (the source of which is the mind) and intuition (the source of which is the heart) both occupy a necessary place in the life of man. This problem of the place of the head and the heart in the scale of values had also stirred the imagination of Western thinkers and poets.

The poems Iqbal composed during his three year's stay in Europe revealed a few new contours of his developing ideas. First, his abhorence towards the glamorous and materially prosperous West, which intoxicates rather than enlightens:

> The European tavern keeper's wine brings jubilee,
> It has not sorrow's ecstasy: give home brewed wine to me.

Second, Iqbal started thinking philosophically about the nature of the Islamic community in general and Indian Muslims in particular. With the widening of his intellectual horizon, India and its problems no longer loomed so large before him. Away from the scene of turmoil, he could think more clearly about fundamental issues which had been obscured by the dust of constant controversy. In 1908 he wrote:

> The life of the individual is contingent,
> the life of the community is real,
> Sacrifice yourself for your community,
> burn the magic of the illusory.[15]

From Nationalism to the Conception of Islamic Community (1908-1938)

From 1908 to 1938, Iqbal wrote his major poetical works and his *Lectures on the Reconstruction of Religious Thought in Islam.* When Iqbal returned from Europe, a number of his most significant ideas had already emerged, but these had yet to be developed into a consistent and dynamic philosophy of life, a philosophy which satisfied the intellect and also gave sustenance to the spirit and an impetus to action. The most burning questions of the day in India were concerned, directly or indirectly, with politics, and a man of Iqbal's wide sympathies and interests could not remain indifferent or aloof. Writing in 1920 on Iqbal's poetry, E.M. Forster had observed: "Poets in India cannot be parted from politics. Would that they could! But there is no hope in the present circumstances, one could as easily part Dante from Florence".[16] It is true that Iqbal cannot be separated from politics, but this is so because politics cannot be isolated from human activity. Owing to Iqbal's great influence on the political history of India of that time, it is easy enough to think of him as a political poet, sometimes even as a politician. He was a political thinker, but it is misleading to stress this too much, for it may lead one to think that he was only, or even primarily, a political philosopher for his poetry was the way he enpresed himself.

It would not be true to say that on his return from Europe Iqbal ceased entirely to try to unify India and began to concentrate exclusively on the effort to present in practicable shape his moral and political ideal to the Muslims. He still considered unity a good thing, though he realized that conditions being what they were, his ideal of unity could not be achieved in India. In 1927, Iqbal said in the Punjab Legislative Assembly: "I do not know whether it is desirable to become a nation".[17] Iqbal was concerned with the fate of India as a whole, and especially with the place of the Muslims in India, till the end of his days. Although he wished to see peace and good will between the various peoples in India, he could no longer think of an Indian nation, "in view of India's infinite variety in climates, races, languages, creeds and social systems".[18]

Soon after his return to India, Iqbal wrote: "Islam is something more than a creed, it is also a community, a nation. The membership of Islam as a community is not determined by birth, locality or naturalisation. The expression 'Indian Muhammedan,' however convenient it may be, is a contradiction in terms; since Islam in its essence is above all conditions of time and space. Nationality with us is a pure idea, it has no geographical basis. But in as much as the average man demands a material center of nationality, the Muslim looks for it in the holy town of Mecca, so that the basis of Muslim nationality combines the real and the ideal, the concrete and the abstract." The transition from India to Hejaz has been made, and Iqbal has arrived, quite conclusively, at what is perhaps the most significant idea in his political philosophy—the extra territorial, supranational character of the Islamic community.[19]

The first corollary of this idea is the rejection of a social order which is subject to the limitations of territory or is nationalistic in outlook. But Iqbal was loath to reject anything outright. He recognized that "the idea of nationality is certainly a healthy factor in the growth of communities".[20] He also believed that it was not always necessary to object to nationalism, for it was not always inconsistent or incompatible with Islam.

It comes into conflict with Islam only when it begins to play the role of a political concept and claims to be a principle of human solidarity demanding that Islam should recede to the background of a mere private opinion and cease to be a living factor in the national life. In Turkey, Persia, Egypt and other Muslim countries it will never become a problem. In these countries the Muslims constitute an overwhelming majority and their minorities, i.e., Jews, Christians, and Zoroastrians, according to the law of Islam, are either "People of the Book" or "like the People of the Book," with whom the law of Islam allows free alliances. It becomes a problem for the Muslims only in countries where they happen to be in a minority, and nationalism demands their complete self-effacement. In majority of countries, Islam accommodates nationalism, for their Islam and nationalism are practically identical, in minority countries it is justified in seeking self-determination as a cultural unit. In either case, it is thoroughly consistent with itself.[21]

Why did Iqbal move away from nationalism? Certainly, the ravages of fierce nationalism he had witnessed in the West influenced his judgment. Nationalism was considered by Iqbal to be a "subtle form of idolatry; a deification of a material object," and consequently, "what was to be demolished by Islam could not be made the very principle of its structure as a political community".[22]

Iqbal also saw nationalism as a weapon of European imperialism: "I have been repudiating the concept of nationalism since the time when it was well known in India and the Muslim world. At the very start it had become clear to me from the writings of European authors that the imperialistic designs of Europe were in great need of this effective weapon—the propagation of the European conception of nationalism in Muslim countries—to shatter the religious unity of Islam in pieces".[23] It is suggested sometimes that Iqbal repudiated nationalism because nationalism was a Western concept, and that he was opposed to all things Western. This is palpably untrue; he has given many reasons for his attitude, and there is no reason to doubt them.

The narrowness of the political concept of nationalism was Iqbal's greatest difficulty in accepting it. "From nationalism thoughts naturally turn more towards the idea that mankind has been so sharply divided into nations that it is impossible to bring about unity among them."[24] Such a concept could not but be opposed to the universal nature of his vision. Why did Iqbal move away from nationalism and toward the conception of the Islamic community? The psychological and religious orientation of Iqbal's early education offers a partial explanation, but there are other, perhaps far more important, reasons to be taken into account.

In order to appreciate Iqbal's dedication to the Islamic community, one must first grasp the concept of the *millat* (Islamic community). Iqbal repeatedly stressed that it was not a narrow concept: "It is not the unity of language or country or the identity of economic interest that constitutes the basic principles of our nationality. It is because we all believe in a certain view of the universe and participate in the same historical tradition that we are members of the society founded by the Prophet of Islam. Islam abhors all material limitations, and bases its nationality on a purely abstract idea objectified in a potentially expansive group of concrete personalities. It is not dependent for its life principle on the character and genius of a particular people. In its essence, it is non-temporal, non-spatial".[25]

The second characteristic of the *millat* was its homogeneity. Ideally, Islam was a great unifying force; the unity it could achieve would be a "true and living" unity. The law of Islam does not recognise the apparent natural differences of race, nor the historical differences of nationality. The political ideal of Islam consists in the creation of a people born of a free fusion of all races and nationalities. Nationality with Islam is not the highest limit of political development; for the general principles of the law of Islam rest on human nature, not on the peculiarities of particular people. The inner cohesion of such a nation would consist not in ethnic or geographic unity, not in the unity of language or social tradition, but in the unity of the religious and political ideal; or in the psychological fact of "like mindedness".[26]

For Iqbal, "like-mindedness" was a necessary condition for the fulfillment of his ideal. The lack of this "like-mindedness," implying the absence of unanimity of aims and purposes, as well as of mutual faith and goodwill, made Iqbal doubt the ideal of a nationhood for India. In 1909, he had written: "I have myself been of the view that religious differences should disappear from this country [India], and even now act on this principle in my private life. But now I think that the preservation of their separate national entities is desirable for both the Hindus and Muslims. The vision of a common nationhood for India has a poetic appeal, but looking to the present conditions and the unconscious trends of the two communities, appears incapable of fulfillment".[27] In 1927, he repeated the same thought: "The talk of a united nationalism is futile and will perhaps remain so for a long time to come. The word has existed on the lips of the people of this country for the last fifty years, and like a hen it has cackled a great deal without laying a single egg ... in this country one community is always aiming at the destruction of the other community.... the present state of things is such that the communities do not trust each other; they have no faith in each other."[28]

Consequently, Iqbal asked himself an important rhetorical question: "Look at the history of mankind, it is an unending succession of deadly combats, blood feuds, and internecine wars. Now the question arises as to whether, in these circumstances, it is possible to bring forth a community, the basis of whose collective life will be peace and goodwill".[29] Iqbal had come to believe that the Islamic *millat* was such a political community. Iqbal adopted the conception of the Islamic community as a gradual unveiling of his vision because "the ideal territory for this nation would be the whole earth," but some critics believed it made his philosophy narrower in scope. For instance, Dickinson, in reviewing Iqbal's *Asrar-i-Khudi* (The Secrets of the Self), said: "Thus, while Mr. Iqbal's philosophy is universal, his application of it is particular and exclusive. Only Muslims are worthy of the Kingdom. The rest of the world is either to be absorbed or excluded".[30]

Iqbal answered thus: "The humanitarian ideal is always universal in poetry and philosophy, but if you make it an effective ideal and work it out in actual life you must start, not with poets and philosophers, but with a society exclusive in the sense of having a creed and well-defined outline, but ever enlarging its limits by example and persuasion. Such a society, according to my belief, is Islam. This society has so far proved itself a more successful opponent of the race-idea which is probably the hardest barrier in the way of the humanitarian ideal ... it is in view of practical and not patriotic considerations ... that I am compelled to start with a specific society [e.g., Islam] which, among the societies of the world, happens to be the only one suitable to my purpose. All men and not Muslims alone, are meant for the Kingdom of God on earth, provided they say goodbye to their idols of race and nationality and treat one another as personalities. The object of my Persian poems is not to make out a case for Islam: my aim is simply to discover a universal social reconstruction, and in this endeavour, I find it philosophically impossible to ignore a social system which exists with the express object of doing away with all the distinctions of caste, rank and race, and which, while keeping a watchful eye on the affairs of the world, fosters a spirit of unworldliness so absolutely essential to man in his relation with his neighbours."[31]

Iqbal referred repeatedly to the universality of his ideal. The question then arises: If Islam is extraterritorial and supranational, why was it not sufficient for Iqbal to have exposed the limitations of the political concept of nationalism? Why did he work so hard to win points of material political importance for the Indian Muslims? Did his policy differ from his theory? How, in the final analysis, is it possible for his antinationalism to be compatible with the nationalist movement of which he was a pioneer which resulted in Pakistan? In view of Iqbal's ideal it would not be difficult to answer these questions. In a word: having willed the end, he had to will the means to achieve the end.

In practical terms Iqbal interpreted the unity of man's spiritual and material life as a double edged intellectual weapon. He asserted that the *Mullas* who considered themselves responsible for the Muslims' spiritual welfare, could not justifiably condemn material progress as profane. And those who were concerned with the temporal side of life could not isolate it from the spiritual aspect. Material progress, although necessary, could not be regarded as sufficient in itself. Standing between the two world views, Iqbal maintained:

> Body lives and spirit lives
> By the life their union gives.[32]

If it had been possible to separate the spirit from the body, then it would have been possible to separate the church from the state. A British Catholic, for example, could owe spiritual allegiance to the Pope and temporal allegiance to the Queen of England. But in Islam, the religious order could not be separated from the temporal order. To believe that "religion is the private affair of the individual, and has nothing to do with what is called man's temporal life," was to deny that "spirit and matter, Church and State, are organic to each other."[33] Islam as an extra territorial and supranational polity was concerned with the total life of man and could not confine itself to its spiritual aspect alone. In Iqbal's words, it was an "individual experience creative of a social order. Its immediate outcome is the fundamentals of a policy with implicit legal concepts whose civic significance cannot be belittled merely because their origin is revelational. The religious ideal of Islam, therefore, is organically related to the social order in which it was created. The rejection of one will eventually involve the rejection of the other".[34]

For Iqbal, "the ideal nation does already exist in germ," but the germ had to grow. The incipient nation needed political guidance and intellectual direction. Iqbal provided this leadership, employing the philosophic concept of the individual ego and the collective ego of the Muslims. "Just as in the individual life the acquisition of gain, protection against injury, determination for action, and appreciation of higher values, are all dependent upon the gradual development of the ego-consciousness, its continuity, enhancement and consolidation, similarly the secret of the life of nations and people depends on the same process, which can be described as the development, preservation and consolidation of the communal ego".[35]

Under what political and psychological conditions could the collective ego flourish? Iqbal explained that "Muslim society, with its remarkable homogeneity and inner unity, has grown to be what it is under the pressure of the laws and institutions associated with the culture of Islam." In an environment in which the Islamic culture could not develop, the individual would not become 'a living member of the Muslim community'. In order to internalize the values of Islam, the individual must thoroughly assimilate the culture of Islam. The object of the assimilation is to create a uniform mental outlook, a particular way of looking at the world, a certain standpoint from where to judge the value of things which sharply defines our community, and transforms it into a corporate individual giving it a definite purpose and ideal of its own...the mental outfit of the individual belonging to the Muslim community must be mainly formed out of the material which the intellectual energy of his forefather has produced, so that he may feel the continuity of the present with the past and the future.[36]

Obviously, Iqbal felt strongly about preservation of Muslims' cultural identity. He stated unequivocally: "That which really matters is a man's faith, his culture, his historical tradition. These are the things which, in my eyes, are worth living for and dying for, and not the piece of earth with which the spirit of man happens to be temporarily associated." Referring to Nehru's concept of organic nationalism, Iqbal asserted that if Nehru stood for "the fusion of the communities in a biological sense," he could not subscribe to such a notion. This would have destroyed the entity of Muslim collective life in India, completely thwarting the realization of Iqbal's ideal.

In order to create the intellectual climate for the development of the Muslims' collective ego in India, it was necessary to achieve "self-determination as a cultural unit".[37] Political and territorial separation explicit in the concept of self-determination may be sought as an end in itself, or it may be only a means to an end. Political power for Iqbal was a means to an end (the end being the preservation of Muslim identity), not an end in itself. To M.A. Jinnah, he wrote: "It is absolutely necessary to tell the world both inside and outside India, that the economic problem is not the only problem in the country. From the Muslim point of view, the cultural problem is of much greater consequence to most Indian Muslims".[38]

Iqbal stressed that "the Indian Muslim is entitled to full and free development on the lines of his own culture and tradition in his own Indian homeland." Did he believe this right could be exercised in India? From the beginning *Shariat* was always in his mind though he started airing it after he came back from Europe after completing his education. On March 20, 1937, less than a year before his death, Iqbal again wrote to Jinnah: "The enforcement and development of the *Shariat* of Islam is impossible in this country without a free Muslim State or States." Actually, he had publicly stated this idea in the presidential address to the annual session of the Muslim League in 1930: "The life of Islam as a cultural force in this country very largely depends on its centralization in a specified territory." To Iqbal a "consolidated North-West Indian Muslim State" appeared "to be the final destiny of Muslims, at least of North West India." This state was to be part of a confederation as a "Muslim India within India".[40]

Would Iqbal have supported the idea of Pakistan as a sovereign state? It is almost certain that he would have done so. But for him, the idea of a separate state for Muslims, law of India would have alluded the Muslim intelligentsia. The question for him (had he been alive in 1947) would not have been to choose between nationalism or antinationalism. It would have been the preservation of Islamic culture in India. In 1930, he had put forward a suggestion for a Muslim state in India, believing that this end could be achieved. But if his suggestion had proved unworkable, or unacceptable, he would have accepted any alternative means of achieving the desired goal. If the territorial unity of India had appeared detrimental to the cultural identity of the Muslims, undermining their solidarity, Iqbal would have emphatically rejected it. However, the creation of Pakistan would have been for Iqbal only a means to achieving a universal brotherhood of man baptized with love.

Although this interpretation of his ideas sounds hypothetical, even a casual reading of his published views leads one to believe that they are in accord with the overall pattern of his thought. Only four and a half months before his death, Iqbal reiterated these ideas in a New Year message broadcast on January 1, 1938 from the Lahore radio station:

> "National unity is not a very durable force. Only one unity is dependable and that unity is brotherhood of man, which is above race, nationality, colour or language, and so long as this so called democracy, this accursed nationalism and this degraded imperialism are not shattered; so long as men do not demonstrate by their actions that they believe that the whole world is the family of God; so long as distinctions of race, colour and geographical nationalities are not wiped out completely, they will never be able to lead a happy and contented life, and the beautiful ideals of liberty, equality and fraternity will never materialize".[41]

Theocracy

Iqbal was an Islamic revivalist. He wrote in his Lectures: "The ultimate spiritual basis of all life as conceived by Islam, is eternal and reveals itself in variety and change. Only we should not forget, that life is not change, pure and simple. It has within it elements of conservation also. While enjoying his creative activity, and always focusing his energies on the discovery of new vistas of life, man has a feeling of uneasiness in the presence of his own unfoldment.

In his forward movement he cannot help looking back to his past and faces his own inward expansion with a certain amount of fear. The spirit of man, in its forward movement, is restrained by forces which seem to be working in the opposite direction. This is only another way of saying that life moves with the weight of its own past on its back, and that in any view of social change the value and function of the forces of conservatism cannot be lost sight of...No people can afford to reject their past entirely; for it is their past that has made their personal identity".[42]

But Iqbal's was never an absolute revivalism. He did recognize the immense significance of the forces of social stability and conservatism. The adherence to the principles of the Koran would impart this conservative element. But he also acknowledged the validity of the claims of the liberal schools of Moslem jurisprudence to interpret the 'foundational legal principles' in the light of the experience of the jurists and in view of the changed situations of the day.[43] Iqbal accepted a religious solution of the problems of the modern world. He, hence, condemned the godlessness of Nietzische. He was bitter against the implications of atheistic materialism, contained in sonic forms of modern nationalistic creeds and formulas. He adhered to the ideals of the spiritual emancipation of man and the social evolution of humanity on spiritual lines. Hence, the materialism, externalism, and plutocracy of western civilization repelled him. He condemned Machiavelli as a 'messenger of Satan' because he propagated separation of ethics and politics.[44] He wrote "Both nationalism and atheistic socialism, at least in the present state of human adjustments, must draw upon the psychological forces of hate, suspicion and resentment, which tend to impoverish the soul of man, and close up his hidden sources of spiritual energy. Neither the technique of medieval mysticism, nor nationalism, nor atheistic socialism can cure the ills of a despairing humanity. Surely, the present moment is one of great crisis in the history of modern culture. The modern world stands in need of biological renewal. And religion, which in its higher manifestations is neither dogma nor priesthood nor ritual, can alone ethically prepare the modern man for the burden of the great responsibility which the advancement of modern science necessarily involves, and restore to him that attitude of faith which makes him capable of winning his personality here, and retaining it hereafter".[45]

He, therefore, pleads for the resurrection of the Islamic concept of *Faqr* which imparts strength and provides capability for the conquest of evils and passions. Thus, religion, according to Iqbal is a source of progress. It has been a factor for the elevation of individuals and groups. Pure conceptualism lacks the dynamic power of moving people.[46] Hence what is needed is faith in the abiding continuity of historical heritage and a culture founded upon religious principles. These ideals of social and political resurrection have to be rooted in the acceptance of a spiritual cosmology. Iqbal wrote: "It is only by rising to a fresh vision of his origin and future, his whence and whither, that man will eventually triumph over a society motivated by an inhuman competition and a civilization which has lost its spiritual unity by its inner conflict of religious and political values".[47]

Iqbal had a theocratic conception of political power, and advocated the Islamic religious orientation according to which individual religious experience vitally affects the social environment and the civic polity. He was opposed to the separation of the transcendent and the secular or the religious and the profane. He was hostile to the modern secularist approach which considers religion to be a private affair of the individual. To him, it is religion that shapes all phases of life. The modern orientation to politics expressing itself in the concepts of the sovereignty of people and the supremacy of the general will failed to satisfy him. He went back to the Islamic conception of the sovereignty of God expressing itself in the total governance of all the aspects and phases of man's life by the laws of the *Shariat*. The physical world is not a distinct autonomous entity but is to be regarded as a sphere for the materialization of spiritual principles. Hence, politics and administration should also be organized on spiritual bases. In Islam, state, individual and government cannot be discussed in isolation. The various aspects of a man's life social, religious, political and economic—cannot be compartmentalized. "In Islam it is the same reality which appears as church looked at from one point of view and state from another".[48] "Islam," writes Iqbal, "is a single unanalysable reality, which is one or the other as your point of view varies".[49] There are not two distinct domains, spiritual and the profane, but it is the same reality, and the sacredness or the secular nature of an act is to be determined by the mental attitude at the back. "Thus, the Quran considers it necessary to unite religion and state, ethics and politics in a single revelation much in the same way as Plato does in the *Republic*".[50] Islam, in short, represents an ideal of a harmonious organic structure and wants to unite the sacerdotium and the imperium. The Christian duality of the empire *versus* the church is foreign to it. There is no duality of a spiritual world and a temporal world. He believed in the creation of a human commonwealth based on the acceptance of the sovereignty of God. In place of nationalism which separates, he advocated the concept of Islamic humanism. He wrote: "The state, according to Islam, is only an effort to realize the spiritual in a human organization".[51] Thus, Iqbal sponsored theocracy in the sense of exalting the spiritual principle as the basis of political governance but he never subscribed to the cult of the ruler as the vicegerent of God. He found great inspiration in the lines of the Koran: "Say: O God! Owner of Sovereignty. Thou givest power unto whom Thou wilt and Thou withdrawest power from whom Thou pleasest." This theocracy is a neutralization of force and domination.[52]

Social Justice

Mohammad Iqbal ardently exposed the evils of social oppression and social injustice that had engulfed the Muslim society. Iqbal projected among the progressive ranks of his time. The distinguished feature of his works projected an active, dynamic man. In his public and philosophic writings as well as in his poems, Iqbal did not merely grieve over injustice, but zealously called for a revolutionary transformation of life. He was one of those Indian patriots who thought over the ways of developing his Motherland after it threw away the colonial yoke. He tried to formulate and define his social ideals, no matter how utopian they may appear to some people.

First and foremost in elaborating his ideas of social change, Iqbal proceeded from the actual position, that is, liberation of his country from colonial hegemony. During the first period of his creative life, Iqbal, like many other spokesmen of Muslim intelligentsia, saw the attainment of freedom as a distant perspective and a possibility though remote at that time. For him freedom formed part and parcel of the moral and religious development of the human personality. Without this, according to Iqbal, it was neither possible to attain freedom nor to make use of it.

The end of the nineteenth and the beginning of the twentieth centuries saw social upheavals the world over, caused by the advent of imperialism. The ideas of moral perfection, as an integral part of the search for social ideals, played an important part in the philosophy of many outstanding thinkers of the time. Among them were Bertrand Russell in Britain, Leo Tolstoy in Russia, Rabindranath Tagore and Gandhi in India and many more. It should be stressed, however, that Iqbal was among the Indian patriots who considered the rousing of social consciousness to be necessary for the realisation of social justice.[53]

The meaning of Iqbal's poem addressed to the students of the Aligarh Muslim University is characteristic of his poems written during the first decade of this century. In this poem, Iqbal held that a person who loved his motherland must dedicate all his life to the struggle for the liberation of his country.

> "People are sending many messages:
> Mine is different
> When one loves the fatherland
> One's words have a different ring".[54]

The formation of Iqbal's social ideals was greatly influenced by the Great October Socialist Revolution. It is during that period that he developed his views on social justice. He rejected every form of oppression. He was opposed to all kinds of exploitation of the peoples of the East by Europe, of peasants by landlords, of workers by capitalists. Iqbal's writings were saturated by a deep belief in the creative abilities of the common people. He believed that the future depended upon the working classes. The poet did his best to arouse political consciousness in his compatriots to foster their aspiration for freedom. Earlier his appeal had been mostly to the intelligentsia; after the October Revolution in Russia, he addressed both the intelligentsia and the toiling masses, exhorting them to unite. At that very time, he created a poetical image of the stonemason (*kuhkan*) who acted as his spokesman:

> "Let us give new regulations to the magicians and their temples,
> Let us overturn the foundations of strong old taverns."

The poet compared a leader who dissociated himself from the people to a twig broken away from the tree. He wrote:

> "Hold on the tree, the ever living trunk
> And believe your spring, your great people".[55]

To Iqbal mass upheaval brought about by national liberation movement in the colonial East, which was inspired, in turn by the ideas of the Russian Revolution was a revelation. The collection of his poems, *Payam-e-Mashriq*, published in 1923, was dedicated to Amanullah Khan, ruler of Afghanistan, one of the first independent nations of that time. Iqbal considered the victorious struggle of Afghanistan for political independence to be a prelude to the emancipation of other oppressed nations. Iqbal expressed his opinion in the preface to his book. He wrote that the East, the Colonial east, had awakened. Addressing the Western countries, he said, the time when imperialism would be done away with and slaves set free was not a long way off.[56]

At the same time, Iqbal was aware of the obstructions to the way to freedom. He was concerned over the lack of unity among the Indian leaders having different view points who waged the struggle against foreign rule. He never visualised that would put off interests of landlords and peasants, capitalists and workers would be harmonious. On the contrary, he always attempts to proclaim such harmony would put off peasants and workers from their struggle against these vested interests. Any attempts to divert the poor from the struggle for the right by promise of happiness were sarcastically criticised by Iqbal. Iqbal inquired indignantly what right a capitalist had to swallow up the result of a worker's labour and a landlord to suck the blood of the peasant. One will fail to find in Iqbal's writings the call to submit to social injustice. On the contrary, he time and again exhorted people to launch a united struggle against both foreign rule and the exploiting classes within Indian society. According to him, linguistic and religious animosity obstructed the just fight of the people against oppression of any kind. Iqbal stressed the fact that such animosity was spread and exploited by vested interests to serve their own ends. In his poem, 'Capital and Labour', he wrote about it.

Thus Iqbal understood the urgent need to overcome racial, national and religious limitations. This awareness was never far from his mind by him. He stood by it even at the time when he became one of the prominent ideological leaders of 'Muslim nationalism' and a movement for a separate homeland for the Muslims of the subcontinent. He was an opponent of imperialist exploitation and social injustice. Till the very end of his life, Iqbal was ranked among the progressive thinkers of his time.

Iqbal maintained that only a revolutionary changes could bring down the old world order. He created a magnificent poetical image of Revolution and treated it both from philosophical and political standpoints. Iqbal was one of the first among Indian patriots to realise the significance of the great October Socialist Revolution in Russia for the liberation movement in the country. He comprehended very dearly the ideas that brought about the spread of socialism. He could go a long way in solving many of the ailments of the Indian polity. He welcomed the Revolution. He considered it to be the social upheaval that led to the collapse of the old world and the transfer of power from kings and oppressors to the common people.

> Now is the time for establishing a new order,
> Let us wipe the heart's slate clean and begin everything anew
> The imperial crown is tarnished, it has been cast away,
> *Kuhkan* came with a chipping hammer in his hands and demanded for himself the estate of Parvaiz."[57]

He welcomed the new era, the era of working people, when crowns were siezed from the rulers and a *Kuhkan* took over state power. Iqbal considered workers to be a new political force that was capable of doing away with social evils.

The poet welcomed the necessity of the collapse of the old decayed world. He glorified the rise of social consciousness and people's union for the sake of the implementation of social and moral ideals. He wrote in his poem, '*Inquilab*':

> "The hearts are throbbing with a yearning for revolution.
> Perhaps the days of the old world are numbered".[58]

However, his understanding of the ways which would lead to social changes was vague. It was bound to his ideas of the reconstruction of Islam. He approved of the October Revolution but at the same time saw the difference in the various types of revolutions. In Iqbal's view, the Russian revolutionary methods were not appropriate for colonial India, as he felt that social reformation should be a result of moral perfection. Iqbal's treatment of revolution cannot be properly understood, unless his views on the immediate conditions of the Revolution are taken into account. He believed that those transformations should be based on traditional, moral values, religious values coming first. The core of his philosophy was the thesis of moral development of the human personality. This affected his understanding of the Revolution. Iqbal rejected certain methods of the Revolution because he was concerned over its atheistic nature. He was bothered neither by the character of the working people's power nor by the profoundness of socio-economic transformations. To him a spiritual impetus of any revolution was of primary importance. He deemed that to be in accordance with the true essence of Islam. According to Iqbal, reformation in the West preceded revolutions. Similarly, in the East, the reconstruction of Islam should precede revolutionary changes.

All this constituted the basis of Iqbal's teaching of the reconstruction of Islam. His theory transformed further the ideas of Waliullah and other Muslim Ulema who were opposed to fanaticism. The Ulemas' stand was to interpret religion in keeping with the spirit of every age. Iqbal's doctrine of the actual Man-doer was an integral part of his philosophy. Iqbal treated the human personality as that of being capable of changing the social environment, whose existence was estimated in terms of its usefulness to society. This doctrine of his developed further certain ideas of Syed Ahmad Khan and some other enlighteners of the last decade of the nineteenth century. Iqbal stressed the idea of the reconstruction of Islam being utilised for the implementation of social change for the translation of his ideals of social justice into reality. Many of his poetic compositions presented this idea, the philosophical poem *Javed Nama* being a notable example.

In his search for religious and moral ways to achieve revolutionary changes, he did not always oppose evolutionary methods as against revolutionary ones. According to the logic of his argument, evolutionary changes growing out of the principles of reformed Islam in the long run would result in revolutionary shifts. The very path, in his opinion, should bring an ideal social environment which is opposed to capitalism as well as irreligious socialism.

Iqbal's concept of social justice is represented fully in his teaching of the Ideal State. His social Utopia was just an attempt to co-join instinctive social endeavour to the reformed dogmas of Islam. That teaching anticipated in many respects contemporary theories of Muslim socialism mooted by certain representatives of the Muslim middle class in developing countries.

Khalifa Abdul Haki, an outstanding Muslim ideologist, and some other disciples of Iqbal characterise his social Utopia as "Muslim Socialism" based on the concept that "Islam and Socialism supplement each other".

The question how much Iqbal's social philosophy approached socialism has claimed the attention of many scholars of different trends. Wilfred C. Smith, one of the most prominent scholars on the role of Islam in Indian social thought, wrote that "if Iqbal were made the dictator of a Muslim state, he would first make it a socialist state". Nevertheless, Smith as well as Hafiz Malik and some other scholars, have noted the difference of Iqbal's social philosophy from socialism.[59]

These once again show that Iqbal's social Utopia bore signs of duality. These views of Iqbal revealed his attempts to embody his social Utopia in keeping with his religious outlook. Iqbal agreed to social changes as long as they, from his point of view, were in conformity with the reconstruction of the main dogmas of Islam. Iqbal wrote in one of his letters to Ghulam Saiddain: "As for socialism, Islam itself is a kind of socialism from which the Muslim society has so far not derived any benefit."

Sometimes, Iqbal did not consider the Ideal State to be the result of a revolution. In a letter to Jinnah, he wrote: "For Islam the acceptance of social democracy in some suitable form is not a revolution but a return to the original purity of Islam." In some cases, Iqbal viewed the call to the revolution when he propagated the creation of the Ideal State based on the Quranic dogmas of justice. "The *Quran* is not only a book, it is a message of evolution... its teachings provide guidance to man in every age in accordance with his needs, requirements and altered conditions of life".[60]

Iqbal's teaching of the State was an expression of many humanistic, progressive principles. Although he did not understand genuine scientific socialism, his socio-economic views were anti-imperialist, anti-feudal and of an anti-monopoly nature. Speaking of the real essence of the *Quran* being embodied in the Ideal State, Iqbal stressed: "The *Quran* is a message of death for the capitalist and of succour for the destitute. It forbids accumulation of capital, usury and speculation. It abolishes Feudalism in all its forms, because, according to its principles, land is God's estate, Man's status is merely that of a trustee or lessee who can only seek livelihood from land".

The political system of Iqbal's Islamic State also embraces the ideals of socialism. He opposed Western democracy which, according to him, was absolutely unacceptable to the people of Hindustan. All works by Iqbal contain a severe criticism of Western democracy. He qualified this democracy as "a demon of autocracy" under the guise of "the Queen of Freedom".

In opposition to Western democracy, the Ideal State was an "anti-monarch, anti-monopoly state".

Though speaking of a Muslim Ideal State, Iqbal came out against theocracy as such, as well as against the domination of mullahs and Ulema in running the state.

His comprehension of democracy included the acknowledgement of the supreme power of God and the rule of elected deputies of people as God's representatives on earth. In accordance with his concept of the elective character of a state power and the people's involvement in governing the State, Iqbal saw Muslim institutions as *shura* and *ijma*. He deemed them to be the best implementation of the Republican form of rule.

The basis of Iqbal's teaching on the Ideal State and his conception of social justice was his treatment of the principle of *tauchida*, equality before God, and law treated them as proper instruction of material and social wealth and solidarity of Muslims in the struggle for their rights and all round emancipation. He paid special attention to the safeguarding of Muslim womens' rights.

Imperialism

With the intensification of international tensions, Iqbal began to realize the growing danger of a new world war. Even though he was unable to see the menace to European safety in the rise of Hitler, he believed that Italy and Germany had set out on a feverish armaments race. However, he saw the situation in Asia more clearly, noticing, for example, that Japan's longing for Manchuria, which would bring her to the borders of the Soviet Union, was the reason she had attacked China. He condemned each outrage of the imperialist states, busy manufacturing weapons to flood the world with blood:

> If war is to be abandoned it is necessary
> to educate him [the leaders of Europe]
> Whose bloodthirsty talons endanger the world.
> Europe is wholly immersed in armaments,
> For the protection of false splendor and Might.[61]

Iqbal was well aware of the horrors another world war would unleash. In a bitter satire on the First World War written in 1924, he had warned of the prospect of wholesale killing and the bombardment of the people. He realized that a defense against the expansion of the imperialistic states was necessary. Resignation and pacifism were contrary to the spirit of his philosophy; instead he believed in an armed struggle against evil. Since he was accustomed to express his thoughts in the religious terminology of Islam, his call to action was in the name of *jihad* (the struggle for faith against infidelity). He did not think that any European state should be left with the power of warlike expansion:

> We ask the Shaikh, most solicitous about the church,
> Is war evil in the East? It is also evil in the West.
> If such is the law, then is war forbidden
> In Islam and permitted in Europe.[62]

His belief in the necessity of armed resistance to imperialism is associated with consideration of absolute power in general. He pointed out that empires are built with the flesh and bones of innocent people. In order to dramatize this thought, Iqbal chose the form of a dialogue between a pirate and Emperor Alexander the Great. In this poem (*Ayk Bahri Qazzaq Awr Sikander*), Alexander seizes the seafaring marauder and is weighing the nature of the punishment for his crimes. The pirate, however, objects that the emperor with his judgment implicates himself and all his peers, and declares:[63]

> We shed blood alike, robbers are we—
> you on the land and I on the sea.

This abhorrence of brute power led Iqbal to believe that no government based on force can long endure; several poems in *Darb-i-Kalim* depict the meteoric rise and fall of the empire builders.

> Alexander stormed like lightning,
> And we all know that suddenly he died.

or:

> Nadir Shah robbed Delhi of its treasures and a single stroke of the sword put an end to his short history.

or:

There is no longer the Empire of Parvaiz in the world.[64]

Iqbal believed that the common feature of the colonial politics of all imperialist states was the self-assumed role of civilizing the so-called backward countries of Asia and Africa. This, in his mind, was only a pretext for economic supremacy of the European states.

Where a guardian of civilization is needed,
The issue, today, is not difficult.
It is simply that where there are no dice
 or women in tight fitting dresses,
Where it is held an offense to drink wine.

He devoted several poems to the affairs of the countries in the Near East, and supported the idea of the free development of Islamic nations. His poetry opposed the policy of Great Britain, which after a victorious war with Turkey was guaranteed (in the Treaty of Sevres, August 20, 1920) rule over Syria, Iraq, Palestine, and Arabia. The English had assured the Arabs that after eliminating Turkish rule from the Near East, they would modernize their country.[65] Iqbal indicated that the liberation of Arabia from Turkish rule was a mere pretext for colonial domination. He objected also to the British policy of establishing a Jewish state in Palestine only because Jews had lived there during the course of history. He believed that the Arabs had a similar claim to Spain, since their ancestors had governed there.[66] Iqbal stated that the purpose of the annexation of Palestine was not associated with trade considerations but rather with dividing the Arabs, and this would be a serious obstacle to Islam's future world ascendancy.

If the Jews have a right to the land of Palestine,
Why have not the Arabs a right to Spain?
The aim of British imperialism is somewhat other
Than a question of oranges, honey or dates.[67]

Iqbal noted how imperialist politics dealt with different classes and men in order to break their resistance and subordinate their opinions to its aims. He believed that Islam was a powerful dam against social abuse and a protection against bondage. In the poem *Iblis Ka Farman Apany Siyasiy Farzandun Kay Nam* (Injunction of the Devil to His Political Offspring), Iqbal's devil tells the statesmen of the colonial powers that they should annihilate Islam in the lands they wish to subjugate, that they should drive the spirit of Muhammad from the body of the intrepid Muslim and instil European beliefs in place of the religious sentiments of the Arabs. The Muslim Afghans were adduced as an example of resistance against colonial expansion. Never having known British domination, they resolutely fought for their independence and remained free, presumably because they remained loyal to the teachings of the *Quran*.

> The man who famine-racked still fears no death
> Muhammad's spirit from his breast expel!
> With Frankish daydreams fill Arabia's brain
> Islam from Yemen and Hejaz expel!
> Tear from the true believers their traditions
> From Khutan's meadows the musk dear expel!
> Iqbal's breath fans the poppy into flame
> Such minstrels from the flower-garden expel![68]

Democracy

It is often asserted that Iqbal did not believe in democracy. Iqbal did indeed write against democracy, but criticism is not rejection, and it should be remembered that he wrote much in favour of it as well.

It was not democracy itself but "democracy of the West" that Iqbal had difficulty in accepting. Western democracy, in his view, was a cover for far too many injustices. It was, for instance, a weapon in the hands of imperialism and capitalism:

> The democracy of the West is the same old organ,
> Which strikes the selfsame note of Imperialism;

> That which thou regard'st as the fairy Queen of Freedom
> In reality is the demon of autocracy clothed in the garb
> of deception.
> Legislation, reforms, concessions, rights and privileges
> In the materia medica of the West are but sweet narcotics.
> The heated discussions of assemblies
> Are the camouflage of capitalists.[69]

He has also said that "democracy is a form of government in which men are counted and not weighed." It is not surprising that Iqbal viewed with alarm the concept of political democracy. One must bear in mind that he knew the Europe of 1905 to 1908, not a period in which democracy, as we envision it today, had made a great deal of progress. Although he visited England again in 1931 to attend the Round Table Conference, this was too brief a visit for any significant impressions to have been formed—and the depths of the Depression was not a period in which Europe could show itself to its best advantage.

What really worried Iqbal, however, was not democracy as it is thought of in the West today, a vast complex involving every aspect of man's life and organization while, at the same time, preserving his individual liberty. He was worried about an institution so completely political that it ignored the other elements of human personality. His picture of democracy, as he used and understood the word, was imperfect, even for his time although it must be admitted that the meaning of democracy was imperfectly understood by many Westerners and scorned by many others during that period. Iqbal's view of the West was also imperfect. He had not lived in the West long enough really to understand it. He was unable to picture the West apart from imperialism, and he apparently accepted the old shibboleth that the West primarily fosters materialistic qualities, whereas the East primarily fosters spiritualistic qualities, an idea based on the belief that technology supplants religion.

Iqbal's indictment of Western philosophers involved ignoring half of the intellectual life of the West such as that upon which the entire intellectual awakening of New England was based. Only a superficial glance at the writers whose works marked the flowering of New England will demonstrate this—Thoreau, whose own philosophy of the individual later influenced India; Melville, whose greatest work is a philosophical study of the consuming passion of pride and evil; and Hawthorne. None of these writers, in fact, reflects the qualities that led to Iqbal's blanket indictment.[70]

This is not to condemn Iqbal; it is only to observe that his knowledge of the West was far from perfect, and that his vision of the West was, naturally enough, affected by his position, that of a proud Muslim remembering a proud past, and by his times, those of domination by Western powers.

Iqbal believed that man could achieve a closer relationship with God only by developing himself. "Man," he said, "must create his own world." Yet man can develop his highest nature only with love and this is the reason why Iqbal looked with so baleful an eye on the Western scientist. He saw the scientist as the epitome of a materialistic world—a world without love. What he didn't seem to realize was that both the scientist's modern automobile and the Punjab peasant's bullock cart are tangible, material objects; each is simply a device, like the camel saddle, to improve transportation. Love, in Iqbal's sense of a regenerating spirit, can go into the making of either an automobile or a bullock cart. Science, Robert Frost once said, is a "greater enterprise of the spirit into matter".

In the medieval West, when such a concept was not accepted, all kinds of seemingly useless antics were performed in the name of religion. The behaviour of such pillar saints as Simeon Stylites, who existed for thirty years on a small platform sixty feet above the ground, or Saint Abraham, an eastern hermit, who did not wash for decades and whose biographer unwittingly wrote "His face reflected the purity of his soul," became almost the hallmark of the devout man of religion. These are instances of the spirit trying to ignore the material, and thus denying part of its birthright. Muhammad knew better. This is what he most of all wished to avoid; his religious genius recognized that religion without humanity was artifice, and humanity without material things is impossible. It is not so much that Iqbal did not recognize this, as that he realized extreme spirituality was as easy to achieve as extreme materialism. He had the example of the Kharidjites of Islam before him: the oldest Muslim sect, fanatical, puritanical, and terrorist, as well as that of the twentieth century Communists.

The erroneous picture of the scientist described above is somewhat akin to another error, commonly found in both the East and the West, an error represented by the loud lamentations one sometimes hears about the disappearance of craftsmen; people who, living in a world in which machines are regularly designed to tolerances of a millionth of an inch or less, bemoan the loss of the craftsman of the past who worked to a tolerance of a fiftieth of an inch or more. Today's navigators send rockets with split second precision to the moon or around Venus, when only two centuries ago the sailor brought his ship home across a few thousand miles of ocean by dead reckoning—a not very cultivated bit of guesswork. There is something in all men that makes the past seem more attractive, but before one accepts this hypothesis, he must make sure that the present has really been viewed in its proper light.

Even if Iqbal did exaggerate in emphasizing the materialism of the West, he was doing no more than some Western poets themselves had done. And he was fulfilling the primary role of every great poet: insisting that men look upon the present in its own light. He was attacking false standards—all false standards—not only what he saw as the excessive materialism (and that phrase really means irreligiousness of the West, but apathy, timidity, passivity, and obscurantism, in all its forms. Iqbal attacked these just as he supported individual action, courage, confidence, and respect. His assault on false standards was carried out on essentially religious grounds; his criticism was based on Islam as he understood it. And among those whom he felt were guilty of false standards were many of the religious leaders within Islam. One of the significant things about Iqbal's fight to bring true standards to the fore (so far as the West is concerned), is how this Muslim poet, speaking out of his devoutly Muslim background, paralleled the finest thinking that has appeared in the West.

The democracy that Iqbal assailed, to translate his terms into the Western idiom, was incomplete democracy. The institutions of government are not enough, he said, and the world isn't going to change merely by the adoption of some new voting device.

What is the value of democratic elements of government if they only serve to promote the old conditions, if they bring no relief to the poverty and suffering of many? What Iqbal opposed was not democracy, but any society geared to privileged nations. He cried out against individuals who did not fight to improve their society, who resigned themselves to what was and stopped trying to realize their dreams. He spoke out against those who, afraid of a new order, clung rigidly to the old.

Iqbal opposed, too, any society that seemed to him without a religious foundation, and all his life searched for some way by which the benefits of science might be gained without adopting the purely materialistic conception of the universe he associated with science and scientists.

In his sixth lecture on the reconstruction of religious thought in Islam, Iqbal warned that "the ultimate fate of a people does not depend so much on organization as on the worth and power of individual man."[71] He had no liking for monarchy, and in this same lecture proposed a republican form of government; his objection to democracy was not, therefore, based on a dislike for its political institutions, but because:

> Colossal oppression
> Masquerades in the robes
> Of democracy, and with iron
> Feet it tramples down the
> Weak without remorse.

Such oppression needs to be obliterated, and this can only be done with the appearance of good men.

What is a good man? Said Muhammad Iqbal, it is a religious man who acts; it is a man of right principles who believes in them and in himself. He would agree with Thomas Carlyle who said that the good man is he who unconsciously walks continually in well doing. Iqbal's universal theme is strongly humanistic; his concept of the self involves a plea for respect for one's self and respect for one's fellow man, too. Reduced to its barest essentials, this is about as good a definition of democracy, as it is understood in much of the

West today, as one is likely to find: that form of government and society based upon man's respect for man and upon man's confidence in himself. Even today, the forms are occasionally confused, and the institutionalized patterns are taken for the thing itself. The secret ballot, for example, does not insure democracy, although it makes its achievement easier. Nor does the party system insure democracy—America's history of recurring reform Administrations testifies to that. Unless people will trust each other, tolerate each other, no mere political device can assure a working democracy. In this respect, Iqbal has a message that is pertinent for these times, when men often seem reduced to numbers, and statistical analysis by itself often seems an open-sesame to the solution of all the problems of human society.

The basis of Iqbal's thought is religious. There are some who would pose this as his major difference with the idea of secular democracy. But this is a difference more apparent than real. Secularism as it is conceived in the West—or, at least, in much of the West—is not a denial of religion; historically, it is an extension of religion: religious precepts are taken from those who claim a monopoly over them and spread throughout the population. What people can exist except, perhaps, temporarily without a spiritual foundation? The so-called secular state of the United States is alive with spiritual values; is this secularism? Religious principles, in the comprehensive sense of *din*, are taught children all day long. In the words of Lawrence C. Little, "We teach it in arithmetic by accuracy. In language by learning to say what we mean. In history by humanity. In geography by breadth of mind. In handicraft by thoroughness. In astronomy by reverence. On the playground by fair play." Both Iqbal and Emerson would have thoroughly agreed.[72]

Secularism does not mean an aloofness from spiritual values; it means the acceptance of universal spiritual values coupled with a respect for the right of each religious group and to care for their spiritual needs. Thus democracy, too, has a religious base no less real and no less important because it is accepted naturally and no one, ordinarily, shouts about it.

The vision of Muhammad Iqbal is not very different from the ideal of democracy as understood in the West. Sometimes different words are used to express the same meaning; sometimes the dead ends of western life appear to be interpreted as its real significance; but the idea and the spirit seem to be similar.

Was it not Iqbal who said: "It is here and now that men and women must learn to live justly, decently, sincerely; it is here that they must ceaselessly cultivate their manifold inner powers and try to control the tremendous forces of Nature; it is here that they must learn to utilize them, not for domination, but for service, nor for repression and destruction but for creation and enrichment; it is here that, by devotion to the highest ideals, they can become co-partners in God's creative activity and can help in the establishment of "God's Kingdom on Earth".

Socialism

A new factor in the political life of India which arose first in the mid-twenties and influenced the character of the struggle for freedom was the emergence of a class of industrial workers. Simultaneously, there began to arise a new ideology of the working class which became for the first time in India a political factor in the true sense. The influence of socialistic ideas permeated by degrees the youth and leftist groups of the Indian national movement and opened new horizons for them. Reflections of the movement led by this ideology appeared in Iqbal's poetry in the thirties and influenced somewhat his unchanged religious conceptions. He developed a fresh interest in socialism and in the problems of the workers. The achievements of the young Soviet state were already known in India, and in 1933 the Communist Party of India, uniting former Communist groups in the different provinces, was established. It was proclaimed illegal in July 1934, however, and thus could not widely influence the Indian working class.

Iqbal's admiration for Lenin, leader of the October Revolution, is evident in a well known and often cited poem, *Lenin Khuda Kay Hadhur Mein* (Lenin before God), from the collection *Bal-i-Jibril* (Gabriel's Wing). Iqbal in this poem has Lenin express all the grief and wrath of the working people and their indignation over the injustice that rules in the world. In a subsequent poem Allah charges the angels to destroy the homes of the rich, devastate their fields, and improve the lot of godly serfs throughout the world. Lenin's speech before God and Allah's mandate to the angels certainly cannot be described as an appeal for the overthrow of the capitalistic order. But the fact that Lenin, as a representative of a socialistic state in Iqbal's poetry, defends the rights of workers and condemns the deficiencies of capitalism, turns the attention of the reader to the doctrine of socialism.

In the collection *Darb-i-Kalim* Iqbal invokes the spirit of Marx and Lenin to voice disapproval of the exploitative order of Western Europe. In the well known poem *Karl Marx Key Awaz* (The Voice of Karl Marx), he has Marx assail European economists who intentionally conceal the predatory structure of capitalism and exploit science to serve only the interests of the ruling classes.

> Your chessmatch of research and erudition—
> Your comedy of debate and disputation!—
> The world has no more patience left to watch
> This comedy of threadbare speculation.
> What, after all, sapient economists,
> Is to be found in your biblification?
> A comedy of nicely-flowing curves,
> A sort of Barmecidal invitation.
> In the idolatrous shrines of the Occident,
> Its pulpits and its seat of education,
> Greed and its murderous crimes are masked under,
> Your knavish comedy of cerebration.[73]

Although in the mid-thirties he devoted regular attention to socialism, his interest remained centered in his ingrained religious conviction that the economic basis of socialism was identical with the teaching of the Koran. Similarly, he believed that Islam and socialism had the same aim; to safeguard the sustenance of all people. His opinions never substantially changed. However, he confused a scientific social system with a religious charitable activity initially prescribed by Islam but incapable of removing the roots of inequality. In the poem *Ishtrakiyat* (Socialism), he again disclaimed the oft-cited passage from the Koran: "They will ask thee what they shall expend in alms; say 'the surplus' ".

> From the behaviour of the nations it appears
> That the rapid progress of Russia is not without gain.
> Perhaps it demonstrates at this time the truth
> That is hidden in the words, "Say the surplus".[74]

Iqbal demanded for Muslims a special Muslim socialism. His ideas on this subject greatly resembled the ideas of Gandhi who wished to show that India had its peculiar culture and who found in Gandhism a peculiarly Indian national socialism. He described atheism as the chief flaw in the socialistic order of Russia, although he conceded that the Soviet Union had unwittingly accomplished, to a certain extent, the work of God. Iqbal believed that Soviet anti-religion derived merely from the failure and corruption of the Russian Orthodox Church. In the poem *Balshewiyk Rus* (Bolshevik Russia) he declared:

> Those who assumed that the protection of the church
> would bring salvation to them
> Were predestined to be the instruments of the
> destruction of the church.
> Russian atheism was inspired from on high
> To disrupt the old heathen idols of the bigots.[75]

Iqbal conceded that he did not know the economic and social features of a better society for which he longed. In reality he knew neither its details nor its most general features. He simply called for a society which would rest on a foundation of social justice. It did not have to be entirely Islamic. He did not consider whether or not Islam would be capable of becoming a uniting force for a world society nor whether another doctrine would have greater success. Once he said: "If I should become ruler of a Muslim state, I would first and foremost create in it a socialist state".[76]

This wishful pronouncement is long way from social revolution and establishment of the rule of the proletariat. Iqbal was not satisfied with the existing state of society and longed for a change, but despite his frequent invocations of revolution, he did not know how it should come about. The well-known Iqbalian discrepancy between thought and deed showed itself here too. He thought of revolution as an obscure mystic force which emanates from hidden sources and, with a stroke, changes the world for the better. He always spoke of it symbolically. Nowhere is there found the suggestion that revolution is a necessary ingredient of social evolution, nowhere does he explain which social class should bring it about. He anticipated only its approach:

> A happy life exists neither in Europe nor in Asia,
> There the spirit is beaten to death, here the heart.
> Within the hearts an impassioned revolution is born,
> The death of the old world already approaches.[77]

With a view to purging Islamic society of all foreign elements, he objected to induction into Islamic thought any kind of progress secular and democratic originating in Europe and continuously fought against imitation of the European way of life. He opposed even English instruction in Indian middle and high schools, saying that it restricted the creative capabilities of the students, killing their individuality. Iqbal was not able to recognize what was good in European culture, what could advance Indian society and what would be harmful. Because he considered European civilization in general, he repudiated all its components and was prone to believe that its end inevitably was approaching. Rejecting it for insufficient spirituality and too great an emphasis on the material aspects of life, he explained its ineluctable end on moral grounds:

> The culture of Europe pits heart and intuition
>   against one another,
> The soul of such a culture cannot remain virtuous.
> What is not clean in spirit cannot achieve
> A clear conscience, lofty thoughts or good taste.[78]

Khalifah Abdul Hakim explains the factors which might have determined Iqbal's attitude toward socialism: "The *laissez-faire* capitalism of the industrial West had pulverized humanity into hostile national groups, and within every nation too there was class war because the classes of haves and have nots were at loggerheads. His [Iqbal's] own country was predominantly an agricultural country where no industrial proletariat had developed but the conflict of the landlord and the tenant was becoming an acute socio-economic problem. The usurious money lender was even more callous than the landlord. Through usury and the judicial support of the system set up by British in complete disregard of local traditions, the ownership of land was rapidly passing into the hands of money-lending owners who did nothing to improve the soil or the condition of the tiller. Iqbal would welcome a revolution in which the-do-nothing-absentee-landlord, or the usurious money lender, is swept away".[79]

However, the revolution of Iqbal's choice was not along the lines of communistic socialism with its ideal of absolute equality, representing "an unlimited extension of the ideal of the family to the State" and finding its expression in the maxim "from each according to his capacity, to each according to his needs;" it was more along the lines of the socialism which has as its ideal "not a mechanical equality of all members of society, but rather a potential equality in the sense of the maxim of Saint Simon's followers: 'From each according to his capacity, to each according to his merit,' which has as its fundamental tenet not common ownership, but the elimination of all unearned increment".[80]

Iqbal's sympathy for socialism flowed out of his passionate dislike for injustice and despotism. Economic injustice had starved and depraved the body and religious despotism had shackled the spirit. As demonstrated in his poem *Lenin Khuda Kay Hadhur Mein* (Lenin before God), he was one with Lenin when the latter protested against Western civilization:

> What they call commerce is a game of dice:
> For one, profit, for millions swooping death.
> There science, philosophy, scholarship, government
> Preach man's equality and drink man's blood.[81]

Iqbal was also relieved to see the house of God purged of idols:

Unsearchably God's edicts move; who knows
What thoughts are stirring deep in the world mind!
Those are appointed to pull down, who lately
Held it salvation to protect, the priests;
On godless Russia the command descends:
Smite all the Baal and Dagons of the Church![82]

That Russia was godless, Iqbal knew. That she would remain godless he did not anticipate, as he wrote to Sir Francis Younghusband: "The present negative state of Russian mind will not last indefinitely, for no system of society can rest on an atheistic basis."[83] Iqbal attacked "Atheist Socialism" but never socialism, for to him "Bolshevism plus God is almost identical."[84] In a letter to Jinnah he wrote: "If Hinduism accepts social democracy, it must necessarily cease to be Hinduism. For Islam the acceptance of social democracy in some suitable form is not a revolution but a return to the original purity of Islam."[85]

It is sometimes asserted that Iqbal did not know what socialism was. There are of course several interpretations of the doctrine of socialism. Marxist interpretation describes dialectical materialism as scientific socialism. To be a dialectic materialist is "to regard nature as primary, to hold that matter is independently real and that mind develops out of the material and must be explained in physical terms. This view has to be understood as negation of the Hegelian idealism. The reality of thought and other mental phenomena is not denied, only their primacy. Idealism and materialism are treated as being the only possible philosophical positions".[86] There is no indication in Iqbal's ideas, however, "that there is a dichotomy of matter and spirit,"[87] as has been maintained by Professor W.C. Smith. To say that man has spiritual, as well as physical needs, is not necessarily to admit the duality of spirit and matter. Had Iqbal admitted such a duality, he could not logically have claimed that the union of religion with socialism would amount to Islam. (Here Islam is a symbol of man's material and spiritual unity and development.) Obviously such a statement assumes an organic relationship between matter and spirit. Denying the primacy of the material phenomena over the spiritual, Iqbal rejected the Marxist determination of dialectical materialism. Does this rejection amount to lack of knowledge or understanding? It has never been explained.

Smith has rather naively maintained that Iqbal's writings are "throughout tinged socialistically",[88] while saying at the same time that Iqbal did not know what socialism was. Should it be inferred that Iqbal wrote socialistically without any understanding of this doctrine? Since Iqbal wrote on socialistic themes he obviously adopted a brand of socialism. According to *The Encyclopaedia of the Social Sciences*, the six characteristics common to all socialistic ideologies throughout history are: "First, a condemnation of the existing political and social order as unjust; second, an advocacy of a new order consistent with moral values; third, a belief that this ideal is realizable; fourth, a conviction that the immorality of the established order is traceable not to a fixed world order or to the changing nature of man but to corrupt institutions; fifth, a programme of action leading to the ideal through a fundamental remoulding of human nature or of institutions or both; and sixth, a revolutionary will to carry out this programme".[89] All these characteristics are present in Iqbal's socio-political thought.

REFERENCES

1. Vahid, S.A., *Thoughts and Reflections of Iqbal*, Lahore, Mohammad Ashraf, 1964, p. 197.
2. Chopra, H.L., "*India and Iqbal*", Indo-Iranica (Calcutta), XV, June 1961, p. 33.
3. Iqbal, Mohammad, *Bang-i-Dara,* Lahore, Ghulam Ali, 1953, pp. 88-89.
4. *Ibid.*, pp. 62-73.
5. *Ibid.*, pp. 32-66.
6. Shamlo, Latif Mohammad Sherani, *Speeches and Statements of Iqbal*, Lahore, Al-Minor Academy, 1945, p. 169.
7. Iqbal, Mohammad, *Bang-i-Dara, op. cit.*, pp. 19, 82, 87-88.
8. Singh, Iqbal, *The Ardent Pilgrim*, London, Longmans Green, 1951, pp. 24-25.
9. Iqbal, Mohammad, *Bang-i-Dara, op. cit.*, pp. 66.
10. Vahid, S.A., *Thoughts and Reflections of Iqbal*, Lahore *op. cit.*, p. 115.
11. Iqbal, Mohammad, *The Reconstructions of Religious Thought in Islam, op. cit.*, Preface.
12. Iqbal, Mohammad, *Bang-i-Dara, op. cit.*, p. 66.

13. Foster, E.M., "The Poetry of Iqbal", The Atheneum, December 5, 1920 quoted in Malik Hafeez (ed.), *Iqbal Poet and Philosopher, op. cit.*, p. 141.
14. *Ibid.*, p. 142.
15. Iqbal, Mohammad, *Bang-i-Dara, op. cit.*, pp. 118-138.
16. Quoted in Malik Hafeez (ed.), *Iqbal Poet and Philosopher, op. cit.*, p. 144.
17. Shamloo, Latif Mohammad Sherani, *Speeches and Statements of Iqbal, op. cit.*, p. 173.
18. Vahid, S.A., *Thoughts and Reflections of Iqbal, op. cit.*, p. 173.
19. *Ibid.*, p. 51.
20. Malik Hafeez (ed.), *Iqbal Poet and Philosopher, op. cit.*, p. 146.
21. Vahid, S.A., *Thoughts and Reflections of Iqbal, op. cit.*, p. 115.
22. Malik Hafeez (ed.), *Iqbal Poet and Philosopher, op. cit.*, p. 145.
23. Shamloo, Latif Mohammad Sherani, *Speeches and Statements of Iqbal, op. cit.*, p. 224.
24. *Ibid.*, p. 237.
25. Vahid, S.A., *Thoughts and Reflections of Iqbal, Lahore op. cit.*, p. 115.
26. *Ibid.*, p. 60.
27. Malik Hafeez (ed.), *Iqbal Poet and Philosopher, op. cit.*, p. 147.
28. Shamloo, Latif Mohammad Sherani, *Speeches and Statements of Iqbal, op. cit.*, pp. 72-73.
29. Vahid, S.A., *Iqbal His Art and Thought*, Hyderabad, Deccan, 1944, p. 237.
30. Dicinson, *Review of the Secrets of the Self by Mohammad Iqbal*. The Nation, London, 4-12-1920.
31. Vahid, S.A., *Thoughts and Reflections of Iqbal, op. cit.*, p. 115.
32. Malik Hafeez (ed.), *Iqbal Poet and Philosopher, op. cit.*, p. 149.
33. Vahid, S.A., *Thoughts and Reflections of Iqbal, Lahore op. cit.*, p. 115.
34. *Ibid.*, p. 167.
35. *Ibid.*, p. 162.
36. *Ibid.*, pp. 379, 381.
37. Shamloo, Latif Mohammad Sherani, *Speeches and Statements of Iqbal, op. cit.*, p. 210.
38. Malik Hafeez (ed.), *Iqbal Poet and Philosopher, op. cit.*, pp. 150-151.
39. *Ibid.*
40. Vahid, S. A., *Thoughts and Reflections of Iqbal, op. cit.*, p. 115.
41. Shamloo, Latif Mohammad Sherani, *Speeches and Statements of Iqbal, op. cit.*, p. 222.

Theocracy

42. Iqbal Mohammad, *Six Lectures, op. cit.*, p. 232.
43. *Ibid.*, p. 234.
44. Dar, B.A., *A Study in Iqbal: Philosophy, op cit.*, p. 254.
45. Iqbal Mohammad, *The Reconstructions of Religious Thought in Islam, op. cit.*, Chapter VII.
46. Iqbal Mohammad, *Six Lectures, op. cit.*, p. 248.

47. Iqbal Mohammad, *The Reconstructions of Religious Thought in Islam, op. cit.*, Chapter VII.
48. Iqbal Mohammad, *Six Lectures, op. cit.*, p. 216.
49. *Ibid.*
50. *Ibid.*, p. 213.
51. *Ibid.*, p. 217.
52. *Ibid.*

Social Justice

53. Jafri, Ali Sardar, *Iqbal Commemorative Volume, op. cit.*, pp. 144-145.
54. Iqbal, Mohammad, *Bang-i-Dara, op. cit.*, p. 119.
55. Kiernan, V.G., *Poems from Iqbal*, Bombay, Kutub, 1947.
56. Iqbal, Mohammad, *Payam-i-Mashriq*, Lahore, Ghulam Ali, 1958, Preface.
57. Iqbal, Mohammad, *Bang-i-Dara, op. cit.*, pp. 120-122.
58. *Ibid.*
59. Jafri, Ali Sardar, *Iqbal Commemorative Volume, op. cit.*, p. 148.
60. *Ibid.*

Imperialism

61. Iqbal, Mohammad, *Darb-i-Kalim*, Lahore, Ghulam Ali, 1959, p. 22.
62. *Ibid.*, p. 23.
63. Malik Hafeez (ed.), *Iqbal Poet and Philosopher, op. cit.*, p. 163.
64. Iqbal, Mohammad, *Darb-i-Kalim, op. cit.*, pp. 157, 168.
65. *Ibid.*, p. 161.
66. Hitti, Philip, K., *History of the Arabs*, London, Macmillan, 1956, p. 493.
67. Iqbal, Mohammad, *Darb-i-Kalim, op. cit.*, p. 159.
68. Kiernan, V. G., *Poems from Iqbal, op. cit.*, p. 74.
69. Malik Hafeez (ed.), *Iqbal Poet and Philosopher, op. cit.*, p. 155.
70. *Ibid.*, p. 177.
71. Iqbal Mohammad, *Six Lectures, op. cit.*, p. 216, p. 151.
72. Malik Hafeez (ed.), *Iqbal Poet and Philosopher, op. cit.*, p. 182.

Socialism

73. Kiernan, V.G., *Poems from Iqbal, op. cit.*, p. 70.
74. Iqbal, Mohammad, *Darb-i-Kalim, op. cit.*, p. 138.
75. *Ibid.*, p. 143.
76. Malik Hafeez (ed.), *Iqbal Poet and Philosopher, op. cit.*, p. 172.
77. *Ibid.*, p. 69.
78. Iqbal, Mohammad, *Darb-i-Kalim, op. cit.*, p. 139.
79. *Ibid.*, p. 69.
80. Khalifah, Abdul Hakim, *Islam and Communism*, Lahore, Institute of Islamic Culture, 1953, pp. 136, 150.

81. *The Encyclopedia of Social Sciences, Vol. III*, London, Macmillan, 1957, p. 189.
82. Kiernan, V.G., *Poems from Iqbal, op. cit.*, pp. 42-43.
83. *Ibid.*, p. 72.
84. Shamloo, Latif Mohammad Sherani, *Speeches and Statements of Iqbal, op. cit.*, p. 167.
85. *Ibid.*
86. Malik Hafeez (ed.), *Iqbal Poet and Philosopher, op. cit.*, p. 154.
87. *The Concise Encyclopedia of Western Philosophy and Philosophers*, London, Hutchinson, 1960, p. 117.
88. Malik Hafeez (ed.), *Iqbal Poet and Philosopher, op. cit.*, p. 154.
89. *Ibid.*
90. *The Encyclopedia of Social Sciences, op. cit.*, Vol. XII, p. 188.

4

PAN-ISLAMISM

Until 1905, when he went to Europe for the first time to continue his education, Iqbal had devoted his main attention to the passionate preaching of national unity, without which it was impossible, in his words, to realize the wonderful ideals of freedom and independence for India. Iqbal's poetic metaphors [for example, the metaphor of the captive bird India in the poem *Parinde ke Faryad* (The Bird's Lament)] were widely disseminated far beyond the borders of the Punjab. For multilingual India, where differences had been preserved between upper and lower castes and between Hindus and Muslims, Iqbal's poems about the necessity of overcoming these internal differences and strengthening unity among Indians took on a special patriotic ring.

It is enough to recall *Sada'-i Dard* (A Cry of Pain) which literally burst forth from the poet after one of the Hindu-Muslim clashes on the eve of the revolutionary uprising of 1905-1909. In those years, the colonial authorities were preparing a law on separate political representation for the Muslims and the Hindus, a law that ultimately led to a split in the country's national forces. And, at this time, Iqbal was calling on his compatriots to create a *naya shiwala* (new temple) in which religious barriers would disappear and love for mother India would prevail. He created an inspiring poetic image of Hindu Muslim unity: "We shall erase the boundary of blind hostility in order that we may again walk along the same path".[1]

In the first of his poetic and philosophical statements, Iqbal still took positions not of Muslim but of all Indian nationalism and did not call for the isolation of the Muslims. However, the contradictoriness of his world view was manifested in the fact that while fighting to overcome religious barriers, he remained, above all, a Muslim. In many respects, this circumstance determined the future evolution of his views.

Early in life, Iqbal sought substantiation for his nationalist ideals in the religious traditions of Islam. As a philosopher and religious reformer, he was under the strong influence of the ideas of the pantheistic school of Sufism. The religio-philosophical teachings of the Sufis on the role of intuition in the cognition process and on the interrelation between man and God, as well as the poetry of these great mystics of the East, had a great effect on Iqbal's works. Before his trip to Europe, he came under the influence of pantheistic concepts about man's path to unity with God. In accordance with these concepts, the man who desires to achieve this unity must lapse into a state of ecstasy, experience moments of renunciation of everything earthly, and be interested in nothing. However, Iqbal was far from accepting the Sufi philosophy as a whole.

From the very beginning, he had a strong desire to interpret anew the religious traditions of Islam. He was attracted not so much by the religio-philosophical conceptions themselves, which were limited (in his own words) by the conditions under which they were created, as by the approach of the creators of these conceptions to the possibility of a new interpretation of Islamic dogma in connection with the change in historical conditions and in man's concepts of nature and society. It is for this reason that Iqbal linked his "reconstruction of Islam" primarily with the names of those Muslim ideologists who recognized not the letter but the spirit of the Quran and the *Shariat* (Islamic laws), endeavouring to reinterpret the traditions of Islam as applied to their country and their times and to prove their right to such a reinterpretation. The fact that the religio-philosophical conceptions, and political positions of these predecessors, frequently differed substantially from each other had no decisive significance for Iqbal.

Iqbal's trip to Europe (1905-1908) intensified his commitment to Islam as the chief means of achieving his national ideals. At the same time, it led him to break with Sufism and contributed to the growth of his striving toward a reconstruction of Islam that would take into consideration the achievements of modern science and the philosophy of the West.

Kant, Leibniz, Fichte, Bergson, and several other Western philosophers had a serious influence on Iqbal. However, he borrowed from them mostly the concepts that had points of contact with traditional philosophy. Hegel's dialectics were an important element in shaping the ideas that became the basis of Iqbal's reconstruction of Islam.

Although using Western sources, Iqbal borrowed not so much the philosophical conceptions and definitions as such, but rather the method of research, the approach to the various phenomena of nature and society. Iqbal's attitude toward Nietzsche's philosophy is of great interest in this connection. As with Nietzsche, the man personality, "ego", is at the centre of Iqbal's philosophical conception, despite its religious nature. As with Nietzsche's superman, a characteristic of the ego was the striving toward constant action. However, Iqbal evaluated the relationship of man to God and to society from positions contrary to those of Nietzsche, and Nietzscheism as a whole was unacceptable to him.

From the very first steps along the path of the reconstruction of Islam, Iqbal opposed following blindly the ideas of the West. "This new wine will weaken our minds, this light will only intensify the darkness",[2] he wrote about the bourgeois civilization of the West, constantly contrasting it to the "great reforming role of Islam".

Iqbal's philosophical credo was reflected not only in his religio-philosophical tracts and political speeches but also in his poetry. Iqbal himself considered poetry the "halo of true philosophy and genuine knowledge," and his contemporaries called him a poet philosopher.

After Iqbal's return from his trip to Europe (1909), the ideology of Muslim nationalism found ever clearer expression in his religio-philosophical, didactic, and poetic work. At its base was the idea that spiritual unity founded on Islam was the most important integral feature of the national society. Islam emerges in this ideology as a form of national unity and (in essence) absorbs political thought.

The shaping of the ideology of Muslim nationalism preceded the spread of the ideas of pan-Islamism, which asserted the necessity of uniting on the basis of Islam, irrespective of state, territorial, ethnic, and national differences. The contrast contained in pan-Islamism, of religious unity to national community based on territorial, ethnic, and economic bonds, was unacceptable to the bourgeois society evolving in the East. Therefore, the ideology of Muslim nationalism, which reflected the interests of the Muslim bourgeois strata, proceeded from the necessity of reforming Islam in accordance with the historical, ethinic, and economic features of the development of individual Muslim peoples and from the existence of several Muslim nations inhabiting different territory.

This was the feature of Muslim nationalism, clearly reflected in Iqbal's philosophy, that ensured its comparatively wide dissemination in the period of the rise of the national liberation movement of the oppressed peoples of the East. Religious motivation in philosophical and socio-political ideas was not characteristic of the ideology of Muslim nationalism alone. To some extent, it was also a trait of other nationalist trends in the social thought of the rising bourgeois society of the colonial East. This was a way of reacting to colonial oppression, since the eyes of the colonial peoples religion was a symbol of past independence, a guarantee of the possibility and the necessity of preserving their own distinctive path of development. In India, where the ideology of Muslim nationalism was the ideology of a religious and cultural minority, it inevitably harboured from the very beginning the seeds of political separatism.

Ideas of the existence in India of two nations became widespread among the Indian intelligentsia after the First World War. Some ideologists acknowledged the existence of both of these nations, others, like Iqbal, only of one of them. Attempting to refute certain Western theories that only British domination ensured unity in India, many representatives of the intelligentsia regarded Hinduism and Islam as the basic manifestation of a community of Indians, antedating the British conquest and independent of foreign domination. They saw community consolidation as one form of national consolidation. Thence also arose the tendency to endow the religious community with features of a national community. However, the preaching of community consolidation inevitably came into conflict with the ideology of the all-India national liberation movement, with the ideas of patriotic unity of the entire population of colonial India. It was used extensively by the colonial authorities in India to split the national liberation movement. As a result, some former advocates of an uncompromising struggle against foreign domination turned at the end of the 1920s into ideologists for national separations which inevitably led to the establishment of Pakistan.

In many ways this path was characteristic of Iqbal.

In a letter of March 28, 1909, he wrote: "I myself have been of the view that religious differences should disappear from this country [India], and even I act on this principle, in my private life. But now, I think that the preservation of their separate national entities is desirable for the Hindus and Muslims. The vision of the common nationhood in India is a beautiful ideal, and has a poetic appeal, but looking to the present conditions, and the unconscious trends of the two, appears capable of fulfillment". This statement attests to the inception in Iqbal's world view of the idea of an antithesis between Hindus and Muslims as two nations.

Although he affirmed the existence of a separate nation of Muslims of India, Iqbal did not lose his pan-Islamic sympathies. Muslim nationalism has always had a pan-Islamic tinge and has envisaged the necessity of a close bond among Muslim nations and the establishment of definite supranational pan-Islamic forms of unity. On the eve of, and during the First World War, Iqbal directly tied his hopes for liberation from colonial dependency to pan-Islamic "solidarity" and the caliphate forms of the liberation movement. This was reflected, in particular, in his poem *Shama Awr Sha'ir* (The Candle and the Poet), written in 1912. However, it should be remembered that Iqbal never regarded the Muslim community as only a religious community. According to his interpretation, Islam was a philosophy of life, the embodiment of definite historical and cultural traditions, social and legal institutions, and way of living. He never equated Muslim nationalism with religious intolerance on the one hand, and narrow communal outlook on the other. While upholding Muslim nationalism, he remained an ardent preacher of patriotic unity in the struggle against foreign domination. In no period of his activity, did he deem it possible to set an alliance of Muslims with British colonialists against the Hindus.

On the eve of the First World War, Iqbal cooperated closely with Abul Kalam Azad, Muhammad Ali, and other Muslim nationalists who at that time were simultaneously members of both the National Congress and the Muslim League and their left wings. His activity was aimed, in his own words, at prompting Muslims to become involved in the political struggle. This was incompatible with government service, and in 1909 Iqbal was forced to give up his well paid job as a professor at the Government College in Lahore and to engage in private law practice. It was in these years that he came forth with his "message to the people." This message included an exhortation to the people, expressed in verse as well as in essay form, to free themselves from political apathy and to become involved in the all-India liberation movement. The ideas of this message found reflection particularly in his poem *Tulba-i Aligarh College Kay Nam* (To the Students of Aligarh College).[3]

The poet exhorted students to vigorous action, to participation in the all-India liberation movement. However, at the same time, he recognized that they still could not free themselves from the traditions of the past, that they were still not ready for the struggle for self-administration. He realized that the hour of liberation had not yet struck, that the time to cast down the "brick of tradition" had not yet come. He wrote, too, of the preservation of the rule of past tradition and the consequent unpreparedness of Indians for revolutionary methods of struggle in many other poetic, philosophical and didactic works. Nevertheless, his constant refrain that another consciousness would inevitably come, the consciousness of the necessity of liberation, left no doubt as to his own desire to cast down this "brick" more quickly.

Iqbal was never a consistent supporter of mass revolutionary methods of struggle. In politics, as in philosophy, he turned primarily to the reforming role of Islam. The complexity and contradictoriness of his world view was also clearly reflected in his first philosophical poems, *Asrar-i Khudi* (The Secrets of the Self, 1915)[4] and *Ramuz-i-Bekhudi* (The Mysteries of Selflessness, 1918).[5] In these poems he presented, actually for the first time, substantiation of the necessity of combining the traditions of Islam and their reformation in the spirit of the times with an expostulation of the moral and ethical conceptions of Muslim nationalism and his philosophy of the man activist.

During and after the First World War, these philosophical poems enjoyed tremendous success among the Muslims of India. They were written not in Urdu but in Persian, since Iqbal was appealing not only to his compatriots but also to the Muslims of all the oppressed countries of Asia. At first, these poems roused the Muslim intelligentsia of colonial India. This was attested, in particular, by the utterances of Muhammad Ali, to whom Iqbal's poems were brought in prison, where he was confined for anti-British activity. "We soon realized, however," he wrote in his autobiography, "that this was even greater and more enduring than what Iqbal had so far written."[6] The Muslim intelligentsia was attracted above all by the appeals in these works to Muslim traditions in the name of solidarity among Muslims in the liberation struggle and the reformation of Islam in the spirit of the times, a spirit of libertarian ideals and utopian concepts of freedom for the human personality and social equality. "He was the poet of Islam reawakening in India in the twentieth century, and to no man does Muslim India owe a greater debt than to this modest, and retiring barrister of the Punjab," Muhammad Ali wrote.[7] Iqbal became a ruler of the minds of the Muslim intelligentsia of colonial India.

The Reformation of Islam

Iqbal believed that profound socio-economic changes were necessary for the establishment of social justice. At the same time, he made the implementation of these changes, dependent on the moral perfection of man, in which Islam must have a deciding role.

Iqbal's moral and ethical ideals, socio-political ideals, and philosophical views were embodied in his teachings on the reformation of Islam. The basic ideas of these teachings, which sounded forth for the first time in Iqbal's philosophical poems *Asrar-i-Khudi* and *Ramuz-i-Bekhudi* achieved their ultimate formulation at the end of the 1920s in his lectures on Islam, *The Reconstruction of Religious Thought in Islam*[8], as well as in the philosophical poem *Javid Namah* (The Book of Eternity)[9], which Iqbal himself called his swan song. Moreover, whether Iqbal spoke as a philosopher, as an ideologist of Muslim nationalism, or as a politician, the moral, the ethical, and the human side of any problem was always of paramount significance for him.

Iqbal explained the need to reform Islam by citing the progress of the development of social thought, science, and experience. This reformation was to play, in his words, the same role in the development of progressive thought in the East as did Protestantism in the West at a certain stage in history. It was called upon to liberate reason and at the same time to replace, as did Protestantism, a system of universal ethics with "national systems of ethics."[10] "It is the duty of the leaders of the world of Islam today to understand the real meaning of what has happened in Europe and then to move forward with self-control and a clear insight into the ultimate aims of Islam as a social policy" Iqbal wrote in his lecture, "The Principles of Movement in the Structure of Islam".[11]

Thus, Iqbal openly acknowledged that the reformation of Islam would serve socio-political ends. He strove to demonstrate the need to preserve religion and religious forms of world view for substantiation of the true meaning of the existence of the human personality and the ideal society, as well as for the disclosure of objective truths that cannot be penetrated with the aid of scientific analysis.

Iqbal separated philosophy from religion but awarded the palm of superiority to the latter: "Philosophy, no doubt, has jurisdiction to judge religion, but what is to be judged is of such a nature that it will not submit to the jurisdiction of philosophy except on its own terms. ... While sitting in judgment on religion, philosophy cannot give religion an inferior place among its data. Religion is not a departmental affair; it is neither mere thought, nor mere feeling, nor mere action; it is an expression of the whole man. Thus, in the evaluation of religion, philosophy must recognize the central position of religion and has no other alternative but to admit it as something focal in the process of reflective synthesis".[12]

This was one of the reasons why Iqbal presented his philosophical and socio-political ideas as part of his doctrine for the reformation of Islam.

The basis of his religio-philosophical views was the opposition of the idea of "God as a personality—a higher 'I' to the classical Greek Philosophical idea of 'God as substance'."[13] From the idealistic thesis of an eternally living and eternally creating God, Iqbal draws a dialectic conclusion on nature and society, which are in perpetual spasmodic development.

Borrowing from the Ash'arite school the so-called atomistic theory, according to which the universe consists of tiny particles—jawahir—and modernizing it in the light of the teachings of Leibniz, Iqbal reaches the conclusion that God created not the things themselves existing in reality but the atoms of which they consist. Creative activity does not cease. "Fresh atoms are coming into being every moment, the universe is therefore constantly growing. The essence of the atom is independent of its existence. Existence is a quality with which God has endowed atoms. What we call a thing then, is in its essential nature a combination of atomic acts." Iqbal also says: "The atoms possess inseparable or negative qualities. They exist in opposed couples as life and death, motion and rest and possess particularly no duration. . . . Nothing is a stable nature."[14]

Iqbal himself called nature "the clothing of God" and his own philosophy "really pure monotheism," but according to the just observation of the Italian scholar, Alessandro Bausani, one of the outstanding scholars of Iqbal's religious and philosophical views "this pure monotheism completely deprives the forces of nature of their divine character".[15] Iqbal in fact arrives through teleology at an understanding of dialectic movement.

This is also characteristic of his theory of cognition. He criticizes ideas of the world's unknowability. Citing the Quran, he distinguishes three sources of cognition: inner experience, history, and nature. Knowledge is "a marriage of reason and experience". Iqbal acknowledged that individual knowledge of the universe (the ultimate ego) is always relative, but knowledge, in the sense of discursive knowledge, however infinite, cannot therefore be predicted of an ego who knows and, at the same time, forms the ground of the object known.[16] Despite his idealistic conception, Iqbal arrived spontaneously at a dialectic posing the question of relative and absolute truth. His theory of cognition prompted an endeavour by man to know surrounding nature and himself.

Self-Determination for the Muslims in India

Iqbal's philosophical conceptions led him to the conclusion that his ideals of equality and freedom could be embodied only in an Islamic state, and that, consequently, the Muslims of India had no other course but self-determination and the creation of Pakistan. The idea of Pakistan was a logical outgrowth of his concept of Muslim nationalism. It was most fully reflected in the political speeches, letters, and political tracts of his last period of activity (the 1930s).

In the middle of the 1930s, many representatives of the radically inclined Muslim intelligentsia set their course toward the creation of a special Muslim state in India. As a consequence of the peculiarities of historical development in colonial India, and also because of the influence of British policy, the national aspirations of many representatives of the Bengali, Punjabi, and Sindhi bourgeoisie and middle strata, the rivalries among individual groups of the bourgeoisie and among the various groups of the civil service intelligentsia, and the social contradictions between peasants and landowners and between commercial financial elements and the farming classes of Northern India, all ended up along the line of Hindu Muslim contradictions.

One section of the Muslim intelligentsia continued to struggle for the independence of a united India and bound its fate with the National Congress, whereas another section shared a belief in the necessity of Muslim self-determination and supported the Muslim League, the political party heading the movement for the creation of an independent Pakistani state. On the eve of the First World War, Abul Kalam Azad and Muhammad Iqbal, the two most important ideologues of the Muslims of India, were in the same camp of ardent patriots of Indian bards and ideologists for struggle against foreign enslavement; but at the end of the 1930s, Azad took the path of National Congress, and Iqbal chose a separate Muslim path of struggle for independence and supported the political movement headed by the Muslim League.

Nevertheless, it is impossible not to see definite differences between Iqbal's socio-political views and the views of the leadership of the Muslim League. Unlike the leadership of the League, Iqbal always supported a closer tie between the Muslim intelligentsia and the masses, and he always put the struggle for independence first. In many ways, he reflected the doubts and aspirations of the Muslim intelligentsia and all the Muslim middle strata, who at the end of the 1930s saw no clear way of solving the problems of the anti-imperialist struggle and realizing their hopes for social justice.

Iqbal, the philosopher, never had a clear concept of the differences between the religious community and the national community. Although, he did try to distinguish between the concepts of national community (*qawm*) and religious community (*millat*), he was not always consistent here. For example, he considered Indian Muslims a nation. "Indeed the Muslims of India are the only people who can fully be described as a nation in the modern sense of the word," he stated in his famous presidential address at the 1930 session of the Muslim League in Allahabad.[17] However, in his opinion, this did not mean that any religious community constituted a nation. Thus, the Hindus, in his words, "are not a nation, although this is what they are striving for."

The definition of the Muslims of India as a separate nation did not mean, however, that Iqbal acknowledged the complete coincidence of national and religious community. He did not discount the fact that in other countries where Muslims did not constitute a minority, as they did in India, they were able to achieve greater progress by developing "along national, i.e., territorial lines".[18]

At the same time, Iqbal constantly emphasized that the desire to consolidate on a religious basis was characteristic of the peoples professing Islam. This desire was intensified under the influence of the colonial invasions. "Today it is being gradually realized in the countries of Islam and in the shape of what is called Muslim nationalism," he noted. Thus Iqbal himself defined "Muslim nationalism" as the consciousness of a national religious community characteristic of the stage of the national liberation movement of colonial peoples in his time. Recognizing the emergence of this consciousness as logical, he endeavoured on this basis to substantiate the necessity of state self-determination for the Muslims of the Indian subcontinent. "I would like to see the Punjab, North West Frontier Province, Sindh and Baluchistan amalgamated into a single state. Self-government within the British Empire or without the British Empire, and the formation of a consolidated North West Indian Muslim state appears to me to be the final destiny of the Muslims, at least of North West India," he stated at the above mentioned session of the Muslim League.[19]

Iqbal's statements did not give a clear idea of what Pakistan's political status and interrelations with India should be. "The Muslim demand for the creation of a Muslim India within India is . . . perfectly justified," he said in the 1930 presidential address. It was there that he called for strengthening of unity on a federal basis, basing the formation of autonomous regions on "linguistic and economic community and religious unity".[20] These statements make it possible to suggest that Iqbal considered complete division between the Hindu and Muslim states expedient.

Iqbal emphasized many times that he would recognize only an ideology of religio-communal unity that was based on respect for the interests of other communities. The endeavour to reconcile communalism with nationalism was as characteristic of his socio-political views as it was of the social thought of the Muslim intelligentsia as a whole.

"Nationalism in the sense of love of one's country and even readiness to die for its honour is a part of the Muslim faith; it comes into conflict with Islam only when it begins to play the role of a political concept, and claims to be a principle of human solidarity demanding that Islam should recede to the background of a mere private opinion and cease to be a living factor in the national life," Iqbal stressed in a famous letter to Jawaharlal Nehru. "Nationalism was an independent problem for Muslims only in those countries where they were in the minority. He wrote in the same letter, "In countries with a Muslim majority, nationalism and Islam are practically identical, but in countries where Muslims are in the minority, their demand for self-determination as cultural unification is completely justified".[21]

Thus, in Iqbal's interpretation, Muslim nationalism retained a clearly expressed anti-colonialist trend, was permeated with ideas of social justice clothed in religious form, and did not rule out unification of the representatives of various religious communities and varying state formation, based on national religious unity, in the name of common goals.

Despite the inner weaknesses and contradictory nature of Iqbal's world view, he remained, at all stages of his religio-philosophical and political activity and poetic work, the spokesman for the hopes and aspirations of the Muslim intelligentsia of India, a ruler of its thoughts, an ardent patriot of his homeland, an impassioned fighter against colonialism and any form of man's oppression of man, and a champion of freedom and humanism.

REFERENCES

1. Iqbal, Mohammad, *Bang-i-Dara, op. cit.*, p. 88.
2. Ikram, S.M., *Makers of Pakistan and Modern Muslim India*, Lahore, Ashraf, 1970, p. 176.
3. Iqbal, Mohammad, *Bang-i-Dara, op. cit.*, p. 119.
4. Iqbal, Mohammad, *Asrar-i-Khudi, op. cit.*, p. 88.
5. Iqbal, Mohammad, *Ramuz-i-Bekhudi*, Lahore, Gulam Ali, 1918.
6. Ali, Muhammad, *My Life: A Fragment*, Lahore, Ashraf, 1946, p. 167.
7. *Ibid.*, p. 166.
8. Iqbal, Mohammad, *Six Lectures on the Reconstruction on Religious Thought in Islam, op. cit.*, p. 162.
9. Iqbal, Mohammad, *Javid-Namah*, Lahore, Gulam Ali, 1932.
10. Iqbal, Mohammad, *Six Lectures on the Reconstruction on Religious Thought in Islam, op. cit.*, p. 163.
11. *Ibid.*
12. *Ibid.*, p. 2
13. Malik Hafeez, *Iqbal Poet Philosopher of Pakistan, op. cit.*, p. 129.
14. Iqbal, Mohammad, *Six Lectures on the Reconstruction on Religious Thought in Islam, op. cit.*, pp. 68,70.
15. Bausani, Alessandro, *Iqbal Philosophy of Religion and the West.*
16. Iqbal, Mohammad, *Six Lectures on the Reconstruction on Religious Thought in Islam, op. cit.*, p. 78.
17. Malik Hafeez, *Iqbal Poet Philosopher of Pakistan, op. cit.*, p. 134.
18. Pirzada, S.S., *Foundations of Pakistan, All India Muslim League Documents, Vol. II, 1906-1947*, New Delhi, Metropolitical Book Co. 1982, pp. 153-171.
19. *Ibid.*
20. *Ibid.*
21. Vahid, S.A., *Thought and Reflection of Iqbal, op. cit*, pp. 256-90.

5

THE GENESIS OF PAKISTAN TWO-NATION THEORY

The most intelligent analysis of Iqbal's mind in its pilgrimage to the shrines of so many ideals in search of a solution for Muslim India is the one offered by Sir Hamilton Gibb:

"Perhaps, the right way to look at Iqbal, is to see in him one who reflected and put into vivid words the diverse currents of ideas that were agitating the minds of the Indian Muslims. His sensitive poetic temperament mirrored all that impinged upon it the backward looking romanticisin of the liberals, the socialist leaning of the young intellectuals, the longing of the militant Muslim Leaguers for a strong leader to restore the political power of Islam".[1]

In the pan-Islamic phase of his writings, which began in 1908 and continued until his death in 1938, Iqbal dissociated politics from nationalism and tried to correlate it with religion and culture. This also implied the rejection of the modern western concept of the duality of church and state. In his Presidential address at Annual Session of Muslim League at Allahabad, he said, "If you begin with the conception of religion as complete otherworldliness, then what has happened to Christianity in Europe is perfectly natural. The universal ethics of Jesus is displaced by nationalist systems ethics and policy. The conclusion to which Europe is consequently driven is that religion is a private affair of the individual, and has nothing to do with what is called man's temporal life. Islam does not bifurcate the unity of man into an irreconcilable duality of spirit and matter. In Islam God and the universe, spirit and matter, church and state are organic to each other".[2]

The real significance, of the Prophet's *hijra* from Mecca to Medina in 622 lay in the repudiation of the concept of local patriotism. The concept of political society in Islam does not accept geographical regionalism or a common race as valid criteria for defining an ethnic group. The earth belongs to God, and as such is the common habitation of all men. Potentially Muslim people can live in any part of it. In their universal political life in the midst of humanity in general, there can be only two criteria of political grouping in so far as Islam is concerned—Muslims and non-Muslims. All non-Muslims constitute a single community, he argues on the basis of a *hadis* which is probably apocryphal, and are antithetical to the community of Islam.[3] This does not justify a striving for a Muslim imperialism at the expense of others. In fact the success of historical Islam in carving out empires was detrimental to Islam's cultural growth because of irrelevant borrowings from external cultures. Nor does it mean that Muslims are in any sense a superior or a chosen people. In fact, the Muslim community is potentially and not actually the *khayr al-umam* (best among the communities), not by the mere virtue of following the Prophet of Islam, whose prophethood was meant to promulgate freedom, equality, and brotherhood among all mankind, but by the community's own maximum effort to apply the ethical values of the Prophet's teachings, to harnessing the forces of nature.[4]

How is this community of Islam in diaspora all over the world to be held together? Its central focus can be found in the basic Islamic concept the unity of God (*tawhid*); its externalization as a social force is reflected in the brotherhood of Islam, which is based on the Qur'an and the sunna. Its external symbol, or as Iqbal calls it, its 'cognizable centre' (*markaz-i mahsus*) is the *ka'ba* which, he says, is nothing more than the cynosure of the mind and the eye: the circumambulation is not merely going round and round a relic: 'but between us [Muslims] and the *ka'ba* there is a symbolic association of which even Gabriel is unaware'. This was a poetic way of restating the Wali-Allahi emphasis that by attracting Muslims from all parts of the world for *hajj* (pilgrimage) and bringing them together, the *ka'ba* symbolizes the unity of Islamic society.[5]

The state to which the universal Islamic society can belong is so far an unrealized ideal Muslim state. A great deal of effort is devoted throughout the works of Iqbal to the attempt to define this ideal state in terms of modern ideologies. Of these, he rejects modern western democracy as essentially plutocratic and based on racial inequality and the exploitation of the weak. He does not regard Islam and socialism as necessarily mutually exclusive or antagonistic: he regards the Islamic concept of equality and Muslim rejection of racialism as similar to that in socialist theory, and the socialist elimination of monarchical institutions as parallel to Muslim iconoclasm. In explaining the institution of *zakat* as a voluntary super tax, Iqbal sees the possibility of eliminating the accumulation of wealth in the hands of the few.[6]

Having stressed all these common values, Iqbal finally rejects communism as the ideal form of world government in comparison with Islam. His argument is based on a speculative explanation of the Muslim *kalima* (the attestation of faith): 'There is no God but God'. He argues that in the dialectics of existence, the law of a negative thesis and a positive antithesis operates, in order to have at the synthesis of truth. 'No' (*la*), the negative particle in the Muslim attestation of faith is destructive; the conditional 'but' (*illa*), the positive one, is affirmative and constructive. The nature of life and the universe is a movement from *la* to *illa*, from negation to positivism, from denial to affirmation. Welded together in an order of experienced truth, *la* and *illa* constitute the means of stock taking the universe and of its subjugation. The particle *la* in isolation denotes revolution, destruction of false gods, but not value creating construction. Communism has got bogged down in the stage of *la*, in negativeness, in destruction of old injustices as well as of old values, but has failed to emerge so far into the creative stage of *illa*. 'The religion of that God-ignoring prophet [Karl Marx] is based on the equality of all stomachs.' Communism has, therefore, much in common with western imperialism: 'Both are dynamic and restless. Both ignore God and betray man. One does this by revolution, the other by exploitation. Between these two millstones humanity is ground to dust.'[7]

The ideal state is, therefore, to be sought not in any other contemporary system but in Islam itself. Thus, one arrives at the concept of the Qur'anic state which is an ideal so far unrealized in Islamic history, and which must not be confused with the caliphate of the four orthodox Caliphs, as is done by most Muslim revivalists. The Qur'anic state is still unrealized and dormant in the mind and conscience of a Muslim. This Qur'anic state cannot be built upon any regional or racial or group loyalty. It would not accept personal rule, and would equate the concept of the caliphate with the service of humanity. In this ideal state, man's vicegerency of God would fulfil itself and honour its trust, by acknowledging that the ownership of all land (i.e. the means of production), vests in God, and by accepting that man's duty is to produce wealth for the benefit of all humanity.[8]

While this ideal state was Iqbal's dream of the future, his immediate concern was the fate of the Muslims in the slowly emerging pattern of self-government in India. Here he once again modified his theory of 'cognizable centre' to apply to the specific problem of Muslim India: a Muslim majority in the north west and a Muslim diaspora in the rest of the subcontinent. Whereas the *ka'ba* remained the symbolic 'cognizable centre' of the entire Muslim world, a regional centre in a politically defined regional state, was necessary for the survival of Muslims in the subcontinent. The Muslim majority areas thus constituted a politically realizable 'cognizable centre' in relation to the diaspora. This marked a retreat from Iqbal's original position, his idealistic denunciation of regionalism. He had now come to the conclusion that while a universal Muslim caliphate was impracticable in the modern world, the only tangible form the political expression of pan-Islamism could take was that of multi-nationalism, realizing itself in regional national states.[9]

Iqbal first put forward a proposal for the creation of a separate state in India in 1930, in his presidential address to the annual session of the Muslim League. He began by asserting that Islam, regarded as an ethical ideal plus a certain kind of polity by which expression I mean a social structure, regulated by a legal system and animated by a specific ethical ideal has been the chief formative factor in the life history of the Muslims of India. It has furnished the basic emotions and loyalties which gradually unify scattered individuals and groups, and finally transform them into a well-defined people possessing a moral consciousness of their own. . . . Is it possible to retain Islam as an ethical ideal and to reject it as a polity in favour of national polities in which the religious attitude is not permitted to play any part? This question becomes of special importance in India where the Muslims happen to be in a minority.

Iqbal's answer to this question was: "The religious order of Islam is organically related to the social order which it has created. The rejection of the one will eventually involve the rejection of the other. Therefore the construction of a polity on Indian national lines, if it means a displacement of the Islamic principles of solidarity, is simply unthinkable to a Muslim".[10]

In comparing the concept of Islamic solidarity as a social unity, with the modern western concept of a nation, Iqbal had taken his cue from Ernest Renan, who had pointed out that the concept of nationalism was, after all, something new and a product of modern western history. It was unknown in the same sense in ancient history. But what constitutes modern nationalism? Is it the concept of a common race? But ethnographic considerations do not seem to have formed modern nations: '*la France est celtique, iberique, germanique. L'Allemagne est germanique, celtique, et slave*'. A similar race mixture is observable in every European country.[11] Then can a common language make a nation? Language, says Renan, tends to unite people, but does not force them into a common nationality. Britain and the United States of America are not one nation nor have Spanish and Portugese speaking nations of South America have become one with Spain and Portugal, a common language has been unable to sustain the principle of nationhood; Switzerland, which has three languages, constituted itself into a nation long ago. Unlike Iqbal, Renan held that religion alone cannot constitute a nation though he agreed that neither could geography, but both religion and geography played important roles in the formation of nationhoods.

Iqbal thoroughly assimilated Renan's argument in his own political thought, applying it to the Indian situation. Judged by this standard the Hindu-Muslim political complex cannot be defined as constituting a single nationhood. In the past, the two communities has always been devided on religious lines. There was indication at all of any will to merge their identities in the future. On the other hand, by this standard Muslim India constituted a nation by itself in the subcontinental landmass.[12]

According to Iqbal, the emotional and psychological homogeneity which inspires the will to create a nation does not exist in the subcontinent. It might have been achieved if the administrative and social eclecticism of Akbar or the mystic syncretism of Kabir had succeeded in catching the imagination of the vast masses of India. Instead, these experiments shrivelled into syncretisms because of their rejection by the great religious groupings and by the caste structure of Hindu society. The fact of diversity within the Indian nation has therefore to be recognized. Ignoring it can only lead to inner tensions. Its recognition may well lead to co-operation between its two major communities. Hindu solidarity has its cultural affinities with the Buddhist world in East and South East Asia; while Muslim solidarity has its religio-political ties with the Middle East. India is Asia in miniature.

This was a significant pronouncement, unapologetically forecasting the 'two-nation theory' which had been vaguely suggested by Sayyid Ahmad Khan and Mawlana Muhammad Ali earlier and was finally developed by Jinnah. Vague suggestions of the desirability of a separate Indian Muslim state had been made sporadically and casually, but they remained largely isolated. Iqbal was the first explicitly to formulate the theory of the necessity of the creation of such a state:

"The principle of European democracy cannot be applied to India without recognizing the facts of communal groups.... The Muslim demand for the creation of a Muslim India is, therefore, perfectly justified.... I would like to see the Punjab North-West Frontier Province, Sind and Baluchistan amalgamated into a single state. Self-Government within the British Empire, or without the British Empire, the formation of a consolidated North-West Indian Muslim state appears to me to be the final destiny of the Muslims at least of North-West India."[13]

Iqbal makes the distinction between Islam as conceived as the legal basis of the state and theocracy which connotes fanaticism.

A separate Muslim state within the subcontinent would not be a theocracy. It would provide, on the other hand, an opportunity for Islam 'to rid itself of the stamp that Arabian imperialism was forced to give it, to mobilize its law, its education, its culture, and to bring them into closer contact with its original spirit and with the spirit of modern times. This mixture of modernism and fundamentalism which he has in mind makes hardly any provision for a secular state for Muslims, like Turkey, 'One lesson', he continues, 'I have learnt from the history of the Muslims. At critical moments in their history it is Islam that has saved Muslims and not *vice versa*.'[14]

During 1936-37 Iqbal and Jinnah came into very close contact. politically. Their perspective became identical. In the past around the period of World War I and then khilafat. They had certain differences of opinion. Now they had the same vision, a separate nation for the Muslims of India. They differed, however, on one point. The Muslim despare in other parts of India. That could not be resolved till his death. In a series of letters to Jinnah, Iqbal emphasized that the creation of a separate Muslim state was the only feasible solution for the Muslims and for peace in India. The Muslims of North West India and Bengal should ignore Muslim-minority provinces in the interest of Indian Muslims whole. The real problem facing Muslim leadership was that of Muslim poverty and backwardness: 'the Muslim League will have to finally decide whether it will remain a body representing the upper classes of Indian Muslims or Muslim masses.'

Happily there is a solution in the enforcement of the law of Islam and its further development in the light of modern ideas.... If this system of law is properly understood and applied, at least the right to subsistence is secured to everybody.... For Islam the acceptance of social democracy in some suitable form and consistent with the legal principles of Islam is not a revolution but a return to the original purity of Islam."

In his introduction to this series of Iqbal's letters, Jinnah acknowledged that Iqbal's views finally led him to the same conclusions, i.e. the demand for a separate Muslim state, which found due expression in the 'Pakistan' resolution passed by the Muslim League in its annual session in 1940.

REFERENCES

1. Gibb, Hamil, *Modern Trend in Islam*, Chicago, U.P., 1947, p. 61.
2. Pirzada, S.S., *Foundations of Pakistan, All India Muslim League Documents, Vol. II, 1906-1947, op. cit.*, pp. 153-171.
3. Iqbal, Mohammad, *Asrar-i-Khudi, op. cit.*, pp. 69-72.
4. Ahmad, Aziz, *Islamic Modernisation in India and Pakistan*, London, Oxford University Press, 1967, p. 157.
5. *Ibid.*
6. Iqbal, Mohammad, *Payam-i-Mashriq, op. cit.*, p. 244.
7. Iqbal, Mohammad, *Javid-Namah, op. cit.*, p. 69.
8. *Ibid.*, pp. 74-83.
9. Iqbal, Mohammad, *Six Lectures on the Reconstruction on Religious Thought in Islam, op. cit.*, p. 151.
10. Pirzada, S.S., *Foundations of Pakistan, All India Muslim League Documents, Vol. II, 1906-1947, op. cit.*, pp. 153-171.
11. Ahmad, Aziz, *Islamic Modernisation in India and Pakistan, op. cit.*, p. 160.
12. *Ibid.*, p. 161.
13. Pirzada, S.S., *Foundations of Pakistan, All India Muslim League Documents, Vol. II, 1906-1947, op. cit.*, pp. 153-171.
14. *Ibid.*
15. *Ibid.*

6

THE DOCTRINE OF PERSONALITY

Iqbal enjoys wide popularity not only in India and Pakistan but throughout the East, especially among Muslims. However, it should be noted that in evaluating the ideas of this outstanding thinker-poet, one notices several extreme and, in fact, mutually exclusive points of view. Reactionary forces, after stamping out everything in his works that is progressive and advanced, try to use his name and prestige to cover up and justify their own narrow selfish and egotistical interests. Also, there are progressivists and pseudo progressivists who declare Iqbal's views reactionary and conservative on the grounds that he spoke, so they say, only to the Muslims because he derived the arguments for his theories and the awakening of the East from the Qur'an. Finally, there are also admirers of his world view who strive with all their might to prove that Iqbal's ideas are the only true ideas, that they must become the guiding star for mankind and that they very nearly serve as its salvation in our age of general despair and hopelessness.

As far as objectively irreconcilable opinions like these are concerned, the work of Iqbal is not exceptional. It is well known that the world view of nearly all great thinkers in any area of spiritual culture has caused, in different eras, the most varied reaction in various strata of society. But in addition to this general reason for such antithetical evaluations of Iqbal, there is also another serious basis in the very nature of his views, which frequently contain inconsistencies and contradictions, and in the complexity and depth of the development of his creative and spiritual evolution, which reflected the complex and contradictory processes of India's ideological life in the first third of the twentieth century.

Iqbal's conscious activity came at the period in the development of his homeland when it had started to struggle for its liberation from colonial oppression. The basis of this struggle consisted in the requirements of India's social development, hindered by its dependence on British imperialism, which in turn perpetuated the country's feudal backwardness and institutions. But the ideological reflection of the struggle for independence and rebirth, as a rule, took the form of religious ideology in two of its basic and, in India, most widespread forms, Hinduism and Islam.

Nearly all the political and spiritual leaders in India at the time, and indeced in present day India, put forth their nationalistic, patriotic programmes even the most progressive and radical of them—under the banner of loyalty to the traditional religions of their homeland. An analogous picture was and is indeed now to be seen in the majority of the Asian and African countries fighting for their independence. Herein lies the distinctive characteristic of the development of the peoples of the East, as contrasted with the West, where the struggle against feudalism and medievalism was carried on, as a rule, under slogans of atheism, anticlericism, and often even philosophical materialism. As is well known, this circumstance has frequently been depicted in literature as the inherent spirituality of the peoples of the East and as their age old adherence to a traditional religious way of life, which allegedly distinguishes the East from the materialist, utilitarian-pragmatic civilization of the West. It is precisely at this point that we have the beginning of the two concepts of Europocentrism and Orientocentrism, which at first glance seem to contrast but are in essence equally fallacious.[1]

The question of the religiosity and traditionalism of the spiritual life of the countries in the East is a rather complicated but, more importantly, an independent subject; however, since it has a direct bearing on Iqbal's world view, at least a few words must be devoted to it. The religiosity and traditionalism of spiritual life in many of the countries of the East has been conditioned by a whole series of factors, of an objective as well as a subjective nature, among which the following may be singled out.[2]

1. Under the conditions of the Asian and African peoples' socio-economic and cultural backwardness, caused primarily by their long sojourn in the chains of colonial slavery, and in the face of a trampling of their national cultures by foreign powers, the heritage of the past and the traditional religion stand forth almost inevitably as the symbol of national pride and distinction, and as a means of expressing patriotic emotions. Moreover, religion has been connected with and has had attributed to it all the grandeur and achievements of Indian civilization in past ages, which have proved especially dazzling against the background of the arbitrary rule and ruthlessness of the colonial systems. It is for this reason that the source of a future renaissance for the motherland was sought in religion.
2. The extensive influence of religious ideology on spiritual life in the countries of Asia and Africa is also furthered by certain features of the religions most widespread there Islam, Buddhism, Hinduism, and others. As distinguished from the Christianity prevalent in the West, religious creeds in the East are not rigidly centralized, clearly organized religions. They lack a clergy appointed from above, standing above and opposed to laymen, and they are far more intimately intertwined with the daily life of the masses. Therefore, any social concepts in the East are most likely to find access to the popular masses, held captive by religious vestiges and prejudices, by appealing to their religious emotions and sentiments. This is why religion, more often than not, is becoming the mouthpiece of social movements, the reactionary as well as the progressive, including the national liberation movements.
3. The religiosity of the national liberation movement ideology in the countries of Asia and Africa is also determined by the character of its leadership: because of the small numbers and poor organization of the working classes in these countries, leadership is in the hands of the bourgeois and petty bourgeois classes and the non-proletarian

parties. By virtue of their social nature, they themselves are highly susceptible to a religious idealistic world view and for that reason are incapable of resisting the influence of a peasant element enmeshed in religious prejudices.

4. Besides, the national liberation movement and the ideology of nationalism in the countries of the East were in the formative process during the historical period (end of the nineteenth century, beginning of the twentieth) when the capitalist system, reigning supreme in the West, was disclosing clearly its exploitative nature, as well as its internal contradictions, when it had begun to disregard the ideals of democracy and humanism. The colonial peoples, who had become the object of capitalism's exploitation, experienced these sides of it particularly keenly. It is completely natural that the capitalist civilization of the West not only could not be acceptable to the thinkers in the East but, on the contrary, excited the sharpest condemnation among them. Moreover, the anti-humanist nature of capitalism presented itself to them as the consequence of the excessive attention paid in the West to the material side of life, which was supposed to be the result of the "worldly utilitarian" nature of Western civilization in general. It is true that the ideologies of nationalism in the East usually have not rejected the material values of life and have acknowledged them as an important and necessary element of social progress, giving the achievements of the West their due in this respect. But, at the same time, these ideologies have held that in order to avoid negative consequences, the material aspects of life must be subordinated to the spiritual values that allegedly find their highest expression in religion, including, primarily, the religions of the East.

Thus, by virtue of all these and an entire series of other reasons, religious motifs and sentiments burst forth in a mighty stream in the teachings of the overwhelming majority of the ideologists for Afro-Asian nationalisms. It is now important to stress that these reasons bear a concretely historical and socially determined character, and so there are no grounds for talk of the inherent religiosity and spirituality of the peoples of the East.

Of course, religion is far from the most adequate form of expression for the progressive substance of a national liberation movement. Nevertheless, in the majority of the countries of the East, for the reasons noted above, religion has proved to be the ideological banner of processes that in their significance for all mankind are revolutionary. Therefore the slogans "revive Hinduism" and "revive Islam" cannot always be the criteria of how reactionary one thinker or another was or one current or another was, under the specific conditions in India at that time. Everything depends upon which forces backed them and what meaning they attributed to what would seem to have no relationship whatever to practical life, that is, religio-scholastic questions concerning the soul and its relationship to God and the world. When we evaluate the highly important role played by religion in India, these questions take on great significance even in the sphere of the political struggle. The aspect of this problem which interests us most, freed of its religious trappings, boils down ultimately to: how man should behave in society?

Some thought that since the soul had no independent existence but was an illusory manifestation of the only true and genuinely real existence, God, the absolute, etc., the ultimate meaning of life therefore consisted of merging with this absolute, of becoming one with it by renouncing, or at least having an indifferent attitude toward, the worldly vanity of vanities. Posing the question in such a way led to the preaching of asceticism, suppression of the will, and, finally, passive submission to the existing state of affairs. Representatives of the patriotic community opposed this trend. Proceeding from the same religio-idealistic premises, they demonstrated the reality of the existence of the human personality and the human soul, the affirmation and strengthening of which are the only aim of life and purpose of man. Hence the recognition of man's high mission in this world and respect for his individuality and dignity. Such an approach provided the basis for demanding the creation of the proper conditions for developing the personality (freedom first of all), for demanding a change in the existing situation and a struggle to reform it. Such was the atmosphere in which this ideology responded to the aspirations of the progressive forces in Indian society who were interested in delivering the country from colonial dependency, weakening the bonds of feudalism, and giving full range to bourgeois attitudes. A host of representatives of the "Indian Renaissance", including Swami Vivekananda, B.G. Tilak, Aurobindo Ghosh, Rabindranath Tagore, Mahatma Gandhi, and others, preached these views. Among Muslims this role, to a significant degree, fell to Sir Sayyid Ahmad Khan and Muhammad Iqbal.

Iqbal is justly called the poet of philosophers and the philosopher of poets. He thought that art, including poetry, must above all shed light on life's most profound problems instead of wallowing in its trivia. It must strengthen man's creative ability and inspire him to struggle for a better life, not lure him away from this struggle, as the supporters of art for art's sake urge. "The highest art is that which awakens our dormant will force and nerves us to face the trials of life manfully. All that brings drowsiness and makes us shut our eyes to Reality around, on the mastery of which alone Life depends, is a message of decay and death. There should be no opium eating in art. The dogma of art for the sake of art is a clever invention of decadence to cheat us out of life and power", Iqbal wrote in 1916.[3] He preserved and carried this understanding of the high goals and civil calling of art through his whole life.

The Reality of Personality

The main feature of Iqbal's works and world view is the doctrine of personality, the reality of its existence and the necessity to do everything to strengthen it. "The moral and religious ideal of man is not self-negation but self-affirmation and he attains this ideal by becoming more and more individual, more and more unique".[4] This concept stands out in all of Iqbal's works and is his determining, in fact his starting, point in his solutions to all the other cosmological problems.

Close acquaintance with the West leads the poet to the conviction that the main reason for India's backwardness (the backwardness of the entire East, for that matter) and humble position lies in its widespread and dominant systems of philosophy that preach passivity, self-denial, and indifference to the living conditions that surround man. Therefore, Iqbal directs all the enthusiasm of his works against these particular views among the Hindus and Muslim as well as against the European varieties of idealism, to which he contrasts his theory of the soul as an active, creative source, eternally seeking and never stopping in its search. "Obviously, this view of man and the universe is opposed to that of the English neo-Hegelians as well as to all forms of pantheistic Sufism which regard absorption in a universal life or soul as the final aim and salvation of man."[5]

In order to prove the existence of the personality, Iqbal considers the entire universe and the entire material world as a composite of individual egos, isolated and separated from each other. "All life is individual; there is no such thing as universal life".[6]

Iqbal attached great importance to this point—the independence and self-determination of the ego—and considers it the basic characteristic of individuality: "The nature of the ego is such that, in spite of its capacity to respond to other egos, it is self-centered and possesses a private circuit of individuality excluding all egos other than itself".[7] In its higher manifestations, in man and in God, individuality becomes a self-recontained exclusive centre.[8] The personality reveals its existence in its constant formation of desires, passions, and ideals.

> We live by forming
> We glow with the sunbeams of desire![9]

Desires and passions do not arise spontaneously, but under the influence of a yet higher emotion, love, which Iqbal has elevated to the rank of the basic, all determining, universally known moral philosophical category. "This word is used," Iqbal explains, "in a very wide sense and means the desire to assimilate, to absorb. Its highest form is the creation of values and ideas and the endeavour to realize them".[10] Thus, Iqbal does not merely believe that the existence of the personality is confined only to the production of ideals and desires. They are only the initial stage of life, which in its true embodiment must manifest itself in creative activity to subordinate and re-shape the eternal world.

In opposition to the theory that the personality as such has no objective criterion for its existence and that the fact of its so-called self-evident being is based solely on the Cartesian conviction *cogito ergo sum*, Iqbal said that the reality of the individual manifests itself not in the act of contemplation but in his purposeful, practical activity in the world, which is the only thing fashioning the personality and demonstrating its objectivity: "The final act is not an intellectual act, but a vital act which deepens the whole being of the ego, and sharpens his will with the creative assurance that the world is not something to be merely seen or known through concepts, but something to be made and re-made by continuous action".[11]

Thus, genuine individuality proves to be inseparable from activity and the active, creative life. In one of his poems Iqbal symbolizes this idea with the image of a wave, which says of itself:

> When I am rolling, I exist,
> When I rest, I am no more.[12]

Tireless functioning, activity, and the struggle against difficulties are, properly speaking, life itself, which in this manifestation is a synonym for freedom.

For Iqbal, only he who considers himself the master of life can claim the right to call himself a real man, a momin. He is a "true believer," because he deems himself the master of life; otherwise he is an "infidel," a *kafir*. Belonging to different religious faiths has no significance here; it is not what determines the value of the individual:

> He who feels himself lost in the world is a *kafir*.
> But he who knows, he is sovereign of the world a *momin*.[13]

To all intents and purposes, true individuality, according to Iqbal, is characterized primarily by its activity and its active relationship to the world. On this level he deemed it possible to speak of immortality, which is not a right or quality residing in man, but which can and must be acquired in the process of constant action. Iqbal places his ideal of man so high that he proclaims him, in the highest and ultimate form of his development, an earthly god or the deputy of God on earth, a model of the harmonious combination of thought and action, instinct and reason, the model toward which it is necessary to strive and whose attainment by all people will signify the advent of the Kingdom of God on earth.

A state of lassitude and inertia indicates the absence of individuality, tantamount to its death. From this Iqbal also draws the criterion for the ethical values of good and evil: Everything that strengthens the personality is good; everything that weakens it is evil. These are the positions from which religion, ethics, and art must be evaluated. It is from these very positions that Iqbal opposed Plato: the poet perceived in the influence of Plato's teachings on Muslim ideology the main reason for the loss of its original life affirming spirit.

Asking (*swal*) leads to the weakening, humiliation, and degradation of the personality. Under the definition of *swal*, Iqbal says, comes everything that is achieved without personal effort, for example, the position of the rich heir or the status of a man who lives on another's thinking in general, any blessing acquired without fighting for it.[14]

All the sobering and striking significance of these words becomes especially clear if we remember that in India of that period sweet illusions were widespread among the intelligentsia regarding the possibility of persuading the colonial power of the necessity of giving up their domination over their country by peaceful means, through "good behaviour." In the poem *Taskhir-i-Fitrat* written to the motifs of Milton's *Paradise Lost*, Iqbal says (in condemnation of these sentiments), that man can achieve his paradise not as the result of someone else's charity, mediation, or intercession, but only as the result of developing and strengthening his own personality; that is, happiness is acquired through one's own efforts and not through the remission of sins by someone else.[15]

When the question was posed this way, Iqbal could not consider the external world an illusion whose fetters bind the spirit in its drive to merge with God the Absolute. On the contrary, for Iqbal the material external world is a necessary factor of the personality's being. "Nature," he says, "is not evil, since it enables the inner powers of life to unfold themselves."[16]

The Individual in Society

The most intimate and direct reality, in the sphere of which all the qualities and potentialities of the individual are realized and his purpose in life effected, is society. Only here man can unfurl his creative activity, approach the ideal of man, gain freedom, and make himself immortal. The individual personality must take a social path and devote itself to serving society. This does not at all mean the loss of its individuality; on the contrary, the social path enables personality to realize itself

In the light of this, the futility of any comparison, even a parallel, sometimes drawn between Iqbal's ideal man and Nietzsche's *Ubermensch* is obvious.[17] It is true that Iqbal spoke sympathetically of Nietzsche, but in such instances, as a rule, he had in mind the venomous attacks on certain aspects of bourgeois civilization that had attracted him to German philosophy, in particular the attacks on clericalism. Iqbal said of Nietzsche that "he plunged his sword into the heart of Europe and his hands are stained with the blood of the clergy".[18] But basically in their concept of man and his purpose in the world—they are radically different.

In his striving for power, Nietzsche's *Ubermensch* cynically tramples all the generally accepted moral and ethical values, and the people for him are a mere crowd, a herd, above which he must proudly rise and reign. Iqbal, on the contrary, wants the entire people to be made up of strong willed personalities united by the common ideals of friendship, fraternity, and mutual service. In his works, Iqbal repeatedly disassociated himself outright from Nietzsche's cynical aristocratism. However, Iqbal's positive presentation of this question serves as the most convincing argument in favour of the contrast, in principle, of their views.

Only in society does man achieve self-realization, and society, in subordinating man to its social structure, makes him free."[19] Iqbal says that when man dissolves his "I" in society, he finds his true individuality in its past history, past traditions, and future achievements. The individual reflects in himself the diversity of society, and the diversity of society embodies its unity in the individual.

The personality that has entirely devoted itself to society acquires immortality, since this personality, as it were, concentrates within itself the past and future of its dominant social ideas. In other words, man's immortality must be understood not as his personal physical immortality but as the eternity and constancy of human civilization and the human intellect. In devoting himself to the service of society, man thereby makes the good of society the goal of his existence. He merges his limited and finite spiritual world with the world of society, which is immortal in its historical development and infinite in its possibilities, thus partaking of its immortality, infinitude, and omnipotence. Only the social man turns himself into the crown of creation and the master of the universe, which he must take over, subdue, and reshape: "it is the lot of man to share in the deeper aspirations of the universe around him and to shape his own destiny as well as that of the universe, now by adjusting himself to its forces, now by putting the whole of his energy to mould its forces to his own ends and purposes".[20] Once the discussion turns to this aspect of man, Iqbal truly becomes the troubador of man's creative and constructive capabilities, his unlimited possibilities for reshaping and subduing the external world: "Hard his lot and frail his being, like a rose leaf, yet no form of reality is so powerful, so inspiring, and so beautiful as the spirit of man".[21] In his laudation of man, Iqbal places him at least on an equal plane with God the Creator, if not higher. In the anthology *Payam-i-Mashriq* (Message of the East) is the famous poem *Mahawdrah Ma-bayen Khuda wa Insan* (God's Talk with Man), in which man hurls a challenge at the Creator himself.[22]

Personality and the Political Society

What was Iqbal's notion of the interrelation among people in society and its social structure? The main demand on the people in what he considered the ideal society was respect for one another: "The principle of the ego-sustaining deed is respect for the ego in myself as well as in others." In *Javid-Namah* (The Book of Eternity) he states that "humanism means respect for man and realization of the position and place of man".[23]

Unlike many Indian thinkers, Iqbal understood that past history offers no such ideal of society. At the same time, Iqbal believed that Western civilization did not have, nor by its very nature could it have, a society in which relationships among people are based on the principle of friendship and brotherhood with one another. Iqbal's criticism of the bourgeois order is one of the most forceful and impressive sides of his work. In this respect, he has been likened to a man holding a mirror before the West, not unlike Hamlet who became the eternal reproach to his mother's conscience.

During his first visit to Europe, Iqbal developed a life long repugnance toward the atmosphere of general bickering, competition, and pursuit of wealth prevailing in the capitalist world, where everything is bought and sold for money, where spiritual values are converted to ready cash. Iqbal painted realistic pictures of the exploitation and the humiliating position of the common man in the competitive society of the West. In his highly artistic poem *Lenin Khuda Kay Hadhur Mein* (Lenin before God) Iqbal presents Lenin angrily protesting to God the horrors and torments that the worker must suffer on earth under His command.

During the last years of his life, Iqbal demanded equality of property and liquidation of the exploitation and the clericism that maintains the bases for this system. Thus, in the poem Lenin before God (already mentioned), the Creator, having heard Lenin's just protest, orders his angels to destroy injustice in the world and to destroy the palaces of a rich men and establish the rule of the people on earth.

As far as the political organization of society is concerned, Iqbal supported the secular state and was opposed to a state headed by a representative of God on earth who could screen his despotic will behind his supposed infallibility.[24] Specifically, he approved of the republican form of government, since it is, in his words, "not only thoroughly consistent with the spirit of Islam, but has also become necessary in view of new forces that are set free in the world of Islam."[25] Regarding the social bases of the ideal society, Iqbal was undoubtedly under the influence of the successful realization of the concepts of socialism in the Soviet Union. In the poem *Javid-Namah* Iqbal places in the mouth of Jamal-ud-Din Afghani a message of greeting to the Soviet people, who are putting into practice the ideals that correspond to the ideals of Islam—the annihilation of exploitation of man by man and the liquidation of the autocracy of the church and the clergy. "What is the Qur'an? A Message of death to rich men and the weapon of the oppressed." Afghani declares to the Soviet people that "the All-High has charged you to observe justice." The experience of the Soviet people, he says, is an inspiring example for all the peoples of the East: "You have fired our hearts and we have been inspired by your experience. The West has become antiquated, but you have taken the true path and you will live eternally."[26]

The Role of Religion in the Development of Total Personality

However, a positive attitude toward the Soviet state does not mean that Iqbal fully understood and accepted the ideas of socialism. It seemed to him that socialist ideology, and materialism in general, had as their ultimate goal only the satisfaction of people's material needs and that they were not at all concerned with the spiritual side of their existence. Thus he believed, for example, that socialism led to the impoverishment of man's spirit and fettered the inner sources of his spiritual energy. "Modern atheistic socialism, which possesses all the fervour of a new religion, has a broader outlook; but having received its philosophical basis from the Hegelians of the left wing it rises in revolt against the very source which could have given it strength and purpose".[27] Socialism, like any atheistic system according to Iqbal, is an incomplete, one sided reflection of the needs of the individual, which are to be satisfied in their entirety and diversity only by the religious ideology of Islam, for it "requires faith in God, not in a throne. And since God is the ultimate spiritual basis for all life, faith in God is in fact elevated to man's faith in his own ideal nature".[28]

This ethical motivation greatly determines the large place religion occupies in Iqbal's specifically metaphysical and generally philosophical views. He maintains that only religious ideology gives full expression to the whole complex of problems connected with man. Therefore, philosophy and science must acknowledge religion's pre-eminence over all other intellectual tools which enable man to apprehend reality. In this cognitive process religion becomes the focus, concentrating within itself the data of science and philosophy and directing them into further research. On closer investigation one realizes that within this scheme Iqbal means by religion nothing more than a world view, a system of general methodological principles, a synthesis of specific branches of knowledge. Certain sciences, Iqbal says, are limited, since they deal only with a specific object of reality, excluding all others from their sphere of consideration. Consequently, they cannot give man a systematized picture of the totality of this reality, but only fragmentary scraps of it. In fact, for Iqbal individual sciences are like vultures that attack the dead body of nature and then fly away, each with its piece of the carrion.

Iqbal asserts that the results of all concrete sciences must be gathered at some single centre and this centre must naturally be religion, since it correlates the data of all diverse types of experiences with the purposeful activity of man and gives the reliable guidance in his research so that his generalizations will be truly general in import. Citing Rumi's language, Iqbal compares a scientist with a hunter who First tracks the musk ox but eventually comes to the conviction that it is far more expedient to follow the smell of the ox's glands than to grope along after its tracks.[29] Iqbal maintains that rational thought must not be considered the highest form of knowledge, but rather religious intuition, which in contrast to rational thought (which masters reality in its static state by parts and degrees), catches reality in motion as an integrated whole. However, Iqbal's intuition does not bear a mystical, irrational character; he places it above so called rational thought only because he means by the latter ordinary superficial-empirical consciousness, based exclusively on the demands of formal logic. Its inadequacy and limitedness are what Iqbal claims intuition is called upon to compensate for; it overcomes creeping empiricism, since by nature it bears a dialectic character. Iqbal says: "In fact, the logical understanding is incapable of seeing this multiplicity as a coherent universe. Its only method is generalization based on resemblances but its generalizations are only fictitious unities which do not affect the reality of concrete things. In its deeper movement, however, thought is capable of reaching an immanent infinite in whose self-unfolding movement the various finite concepts are merely moments. In its essential nature, then, thought is not static; it is dynamic and unfolds its internal infinitude in time like the seed which, from the very beginning, carries within itself the organic unity of the tree as a present fact."[30]

Present here is the Hegelian concept of thought as a process that its functioning dialectically unfurls from within itself its own definitions, which prove to be, in the final analysis, ontological categories of existence.

Intuitions and Personality

Although Iqbal does not separate the highest form of consciousness, intuition, from ordinary experience and scientific thinking with an impassable wall, intuition is nevertheless qualitatively different from them in one respect: it brings man into direct contact with the eternal, intransient basis of existence, inaccessible to the senses or the rational mind because they focus their attention only on its temporary, fluctuating, changing aspects. Intuition opens to man a conception of ultimate reality whereby "diverse stimuli merge into one another and form a single unanalysable unity in which the ordinary distinction of subject and object does not exist."

Thus, for Iqbal the main determinant of ultimate reality is the unity of subject and object, of thinking and existence: "The consciousness or the idea is not something alien, apart from the original nature of things; it is their ultimate basis and forms the very essence of their existence, entering into them at the very beginning of their existence and stimulating their progressive development toward an independent self-determined goal".[31] This unity and identity of matter and spirit, with which the development of reality logically begins and ends, Iqbal calls the supreme soul, God, the absolute ego.

Nature, or the non-I, is organically related to the supreme soul; nature is not simply a mass of pure materiality located in a void but a definite structure of events, a systematized image of their behaviour. Nature is to the divine soul what the body is to the human soul it plays the role of accumulator of its actions. In other words, Iqbal puts forth a pantheistic interpretation of the question of the interrelation between spirit and matter, between nature and God. The absolute spirit is all of reality, Iqbal says; it is immanent in nature. By knowing nature, one knows the activity of God. Consciousness of this, according to Iqbal, also imparts to the natural sciences that higher meaning in which, by themselves, they are deficient.

Iqbal understands the universe not as "a ready-made product that left the hands of the creator ages ago, but one that has just been laid out in space, like an inert mass of matter upon which time exerts no influence".[32] He believes that in the process of development the world will expand quantitatively—"This is a growing Universe," he says. But in its development the world is not moving toward any pre-established goal, since teleology deprives the world process of its distinctiveness and creative character. "A temporal process," Iqbal wrote, "Must not be regarded as a line already drawn. It is a line that is in a process of developing, the actualization of discovered possibilities. It is purposive only in the sense that it has a selective nature and develops up to its given state by means of active retention and maintenance of the past."[33]

Iqbal considered the ethical principle the chief argument against teleology: a world regarded as a process that is realizing a pre-set purpose cannot be a world of free subjects responsible for their actions. In this case, it would be a kind of stage upon which puppets act, obeying strings being pulled by someone off stage. One can speak of teleology only in the sense that in the process of life's development we constantly work out new goals, tasks, and ideal values, which condition our activity.

The creative activity of the ultimate reality, the supreme soul, God, manifests itself in the formation of individual souls, egos that differ in degree of expression of the personality, the highest and most perfect form of which is achieved in man, who is called upon to become God's partner in His creative activity.

"What is matter? Colonies of egos (individualities) of a lower order from which the ego (individuality) of a higher order rises when their association and interaction achieve a certain coordination".[34] The circumstance, Iqbal says, that the higher arises from the lower does not deprive the higher of its dignity and value. The point is not origin but creative ability, meaning, and vocation. The evolution of life shows that at least in its early stages the physical subdues the spiritual, whereas subsequently the reverse is true: the spiritual subdues the physical and, in the final analysis, acquires complete independence.

Dialectical Contradictions

Thus does Iqbal substantiate in theoretical form his doctrine of human individuality, of which an ethical characterization on a personal and social level has been given above. One cannot help but notice that the theoretical basis of Iqbal's world view suffers from a number of limitations and contradictions.

To begin with, the method chosen by Iqbal of seeking in the Quran the substantiation and sanction for all his propositions cannot be recognized as the most successful, even for the India of that time. However, we must be immediately cautioned against exaggerating the role, significance, and relative weight of religious Islamic ideology in Iqbal's world view. Actually, if we grasp the religious form of his teachings, we will realize that what was significant to him was not so much the religion of Islam as the scientific attitude and humanitarianism of Muslim culture. He maintains, for example, that during the gloomy epoch of the Middle Ages, when the sensual world was declared to have no value, Islam began to preach the reality of the external world, laid the foundations for the modern sciences, and called on people to bring nature under their control and that herein lies its service to world civilization.[35]

In one respect, the religious Muslim setting of Iqbal's views played a great role in the life of India; it assisted, and to some extent, especially in our times, even became the banner for the Indian Muslim elite in substantiating their right to independent territorial existence. As Iqbal Singh writes: "His [Iqbal's] fervent idealization of Islam, not merely as a religion but also as a comprehensive polity, exercised a very far reaching restraining, influence upon the Muslim intelligentsia."[36] Although in his political pronouncements Iqbal subscribed to the thought that the Muslims of India constituted a part of the population separate from the Hindus, this did not prevent him from remaining an ardent patriot of India throughout his life, grieving at her sorrows and rejoicing at her successes; he never preached religious or any other enmity between Hindus and Muslims. Therefore he is equally honoured by both groups, in India as well as in Pakistan.

As for the strictly theoretical bases of Iqbal's views, they suffer from the essentially idealistic premises of his world view. For example, he does not give a convincing explanation of his conception of the ultimate reality, the absolute spirit, a failure which lies in the utter impossibility of demonstrating how the spirit, God, turns out to be the organizing principle of what spirit essentially is not, i.e. nature. His conception of God is contradictory; sometimes he speaks of him as a personality that fashions and directs the world and the behaviour of the individual—a typically theological idea that rules out his own interpretation of God either as a concentrated embodiment of the highest ideal qualities of man or as synonym for the regular organization and orderliness of the world process. This initial idealistic premise to some degree determines the precariousness of the principles of the central element in his teaching of ethics.

Although Iqbal also asserts that the personality finds its individuality only in the social field, in fact he considers it a principle independent from society, both in origin and existence. Thus, he admits of the metaphysical opposition of personality and society, yet fails to see the complex nature of their dialectic unity and difference, when the individual receives his true existence only as the bearer of definite social functions in their economic, political, cultural, and ideological manifestations. By virtue of the specific conditions of the historical development of India, where class antagonism frequently took the form of national, religious, and caste contradictions and hid the essence of social relations, Iqbal did not perceive the class nature of the personality, thereby depriving himself of the sole basis on which this question could be solved. Therefore, his theory of the individual rests on the rather cathetic base of the abstract man torn away from the concrete base of socio-economic relations. Iqbal does not go beyond the limits of speculative and abstract humanism; he contrasts problems of an ethical nature to social themes and sees in moral self perfection the basic problem of educating man. In one of his speeches he said that he who intends to change conditions of existence unfavourable to him must change his mind-set.

In Iqbal's notions of the ideal organization of society one cannot help but note elements of utopianism and liberal amorphousness. Although in general outlines he also opposes exploitation and approves of socialist principles, these principles come closer to petty bourgeois, and in certain features feudal, socialism than to the socialism of Saint Simon, Owen, and Fourier.

In general, it cannot be said that Iqbal left behind a harmonious, consistent system of views; they reflect traces of the influence of many contradictory doctrines. In this connection, Iqbal Singh points out that Iqbal's world view, like those of the majority of the outstanding people in India in the last century, from Ram Mohan Roy and Sayyid Ahmad Khan to Mohandas Gandhi, is not an integral whole but is a sort of mosaic made of different parts, although certain parts of it stand out more than others.[37] Of course, this characterization cannot be recognized as the chief and basic point when we begin to speak of Iqbal, but to some extent it gives an idea of his views.

REFERENCES

1. Malik Hafeez, *Iqbal Poet Philosopher of Pakistan, op. cit.*, p. 265.
2. *Ibid.*, pp. 265-267.
3. Iqbal, Mohammad, *Secrets of the Self, op. cit.*, p. xxiii.
4. *Ibid.*, pp. xviii & xix.
5. *Ibid.*, p. xviii.
6. *Ibid.*, p. xvii.
7. Iqbal, Mohammad, *Six Lectures on the Reconstruction on Religious Thought in Islam, op. cit.*, p. 72.

8. Iqbal, Mohammad, *Secrets of the Self, op. cit.*, p. xix.
9. *Ibid.*, p. xxvii.
10. *Ibid.*, p. xxv.
11. Iqbal, Mohammad, *Six Lectures on the Reconstruction on Religious Thought in Islam, op. cit.*, p. 198.
12. Iqbal, Mohammad, *Payam-i-Mashriq, op. cit.*, p. 198.
13. Iqbal, Mohammad, *Darb-i-Kalim, op. cit.*, p. 39.
14. Iqbal, Mohammad, *Secrets of the Self, op. cit.*, p. 39.
15. Iqbal, Mohammad, *Payam-i-Mashriq, op. cit.*, pp. 97-101.
16. Iqbal, Mohammad, *Secrets of the Self, op. cit.*, p. xx.
17. Malik Hafeez, *Iqbal Poet Philosopher of Pakistan, op. cit.*, p. 273.
18. Iqbal, Mohammad, *Payam-i-Mashriq, op. cit.*, p. 241.
19. Iqbal, Mohammad, *Asrar-i-Ramuz*, Lahore, Ghulam Ali, 1948, p. 99.
20. Iqbal, Mohammad, *Six Lectures on the Reconstruction on Religious Thought in Islam, op. cit.*, pp. 12, 112.
21. *Ibid.*
22. Iqbal, Mohammad, *Payam-i-Mashriq, op. cit.*, p. 132.
23. Iqbal, Mohammad, *Bang-i-Dara, op. cit.*, p. 150.
24. Iqbal, Mohammad, *Six Lectures on the Reconstruction on Religious Thought in Islam, op. cit.*, p. 155.
25. *Ibid.*, p. 157.
26. Iqbal, Mohammad, *Javid-Namah, op. cit.*, pp. 88, 120-122.
27. Iqbal, Mohammad, *Six Lectures on the Reconstruction on Religious Thought in Islam, op. cit.*, p. 188.
28. *Ibid.*, p. 147.
29. *Ibid.*, p. 96.
30. *Ibid.*, p. 6.
31. *Ibid.*, pp. 18-19.
32. *Ibid.*, pp. 55, 106-7.
33. *Ibid.*, p. 55.
34. *Ibid.*, p. 106.
35. *Ibid.*, pp. 14-15.
36. Singh, Iqbal, *The Ardent Pilgrim, op. cit.*, 156.
37. *Ibid.*, p. 233.

BIBLIOGRAPHY

Ahmad, Aziz, *Islamic Modernisation in India and Pakistan*, (London: Oxford University Press), 1967.

Ahmad, Aziz, *Islamic Modernism in India and Pakistan 1857-1964* (London: Oxford University Press), 1967.

Ahmad, Jamil-ud-Din (ed.), *Speeches and Writings of Mr. Jinnah*, Vol. I (Lahore: Ashraf), 1960.

Akbar, Ahmad, *Jinnah, Pakistan and Islamic Identity* (London: Routledge), 1997.

Ahmad, Akbar S., *Jinnah, Pakistan and Islamic Identity* (Karachi: Oxford University Press).

Ahmad, Rizwan (ed.), *The Quaid-e-Azam Papers* (Karachi: East & West Publishing Co.), 1976.

Ahmad, Rizwan (ed.), *Iqbal's Letters to Quaid-e-Azam* (Lahore: Friends Educational Service), 1966.

Ali, Muhammad, *My Life: A Fragment*, Lahore, Ashraf, 1946.

Ali, Sayyid Amir, *A History of the Saracens*, London, Macmillan, 1899.

Ali, Sayyid Amir, *The Spirit of Islam*, London, Christopher, 1922.

Almeida, Prakash, *Jinnah Man of Destiny* (Delhi: Kalpaz Publications), 2001.

Ambedkar, B.R., *Pakistan or the Partition of India* (Bombay: Thacker & Co.), 1946.

Ambedkar, B.R., *What Congress and Gandhi Have Done to the Untouchables* (Bombay: Thacker & Co.), 1946.

Azad, Maulana Abul Kalam, *India Wins Freedom* (Calcutta: Longmans), 1959.

Aziz, Beg, *Jinnah and His Time* (Islamabad: Babur and Amer Publications), 1986.

Aziz, K.K., *Rahmat Ali—A Biography* (Lahore: Vanguard), 1987.

Aziz, K.K., *The Making of Pakistan* (London: Chatto & Windus), 1967.

Bakshi, Illahi, *With Quiad-e-Azam During His Last Days* (Karachi: Quaid-e-Azam Academy), 1978.

Bamford, P.C., *Histories of Non-Cooperation and Khilafat Movements* (New Delhi: Deep and Deep), 1974.

Bausani, Alessandro, *Iqbal Philosophy of Religion and the West*, Pakistan Quarterly, Karchi II, 1953.

Bhutto, Zulfikar Ali, *The Myth of Independence* (London: Oxford University Press), 1969.

Bolitho, Hector, *Jinnah: Creator of Pakistan* (London: John Murray), 1954.

Bose, Subhash Chandra, *The Indian Struggle Part II* (Calcutta: Published for Netaji Publishing Society), 1945.

Buski, Shahid Javed, *Pakistan, The Continuing Struggle for Nationhood* (Boulder Colo. West View Press), 1991.

Chagla, M.C., *Roses in December: An Autobiography* (Bombay: Bharatiya Vidya Bhawan), 1978.

Chandra, Bipin, *Communalism in India* (New Delhi: Vani Educational Books), 1984.

Chandra, Bipin, *India's Struggle for Independence 1857-1947* (New Delhi: Viking), 1988.

Chaudhry, Binayandra Mohan, *Muslim Politics in India* (Calcutta: Orient Longman), 1946.

Chaudhry, Kailash, *Tragedy of Jinnah* (Lahore: Varma Publishing Co.), 1943.

Chaudhary Rahmat Ali, *Pakistan: The Farther Land of Pak Nation* (London: The Pak National Liberal Movement), 1947.

Chopra, H.L., "*India and Iqbal*", *Indo-Iranica* (Calcutta), XV, June 1961.

Cohen, Stephen Philip, *The Idea of Pakistan* (New Delhi: Oxford University Press), 2005.

Collins Larry and Dominique Lapierre, *Freedom at Midnight* (New Delhi: Vikas Publications), 1973.

Constituent Assembly of Pakistan Debates, 11 August 1947, Vol. I, No. 2, pp. 18-20.

Countess Mary of Minto, *India Minto and Morely 1905-1910* (London: Macmillan), 1934.

Coupland, R., *The Indian Problem Part II* (New York: Oxford University Press), 1944.

Cunning John (ed.), *Political India 1832-1932* (London: Oxford University Press), 1932.

Damodar P. Singhal, *Pakistan* (New Jersey: Prentice Hall), 1972.

Dar, B.A., A *Study in Iqbal Philosophy*, Lahore, M. Ashraf, 1944.

Das, Durga, *India from Curzon to Nehru and After* (New York: John Doby), 1970.

Datta, S.K. and Rajiv Sharma, *Pakistan from Jinnah to Jehad* (New Delhi: UBSPD), 2002.

Dutt, R. Palme, *India Today and Tomorrow* (London: Lawrence and Wishart), 1955.

Dwarkadas, Kanji, *India's Fight for Freedom: An Eye Witness Story, 1913-37* (Bombay: Popular Prakashan), 1960.

Embree, Ainslie T., *Imaging India: Essays on Indian History* (London: Oxford University Press), 1989.

Faruqi, Zakir Hussain, *Muslim League Kiyyum*? (Bombay: Maktaha-Sultani), 1947.

Gaba, K.L., *Inside Pakistan* (Delhi: Raj Kamal Publications), 1948.

Gandhi, M.K., *An Autobiography or The Story of My Experiments with Truth* (Ahmedabad: Navajivan Publishing House), 1983.

Dicinson, *Review of the Secrets of the Self by Mohammad Iqbal.* The Nation, London.

Faizee, Atiya Begum, *Iqbal*, Bombay, Academy of Iqbal, 1947.

Foster, E.M., "The Poetry of Iqbal", *The Atheneum*, December 5, 1920

Gibb, Hamil, *Modern Trend in Islam*, Chicago, U.P., 1947.

Gunther, John, *Inside Asia* (New York: Hamilton), 1938.

Hamid, Shahid, *Disastrous Twilight* (Leo-Cooper: Oxford University Press), 1986.

Hassan, Javed, *India: A Study in Profile* (Rawalpindi: Army Book Service), 1990.

Hasan, S.S., *Plain Mr. Jinnah: Selections from Quaid-e-Azam correspondence relating mainly to personal matters* (Karachi: Royal Book Company), 1976.

Hitti, Philip, K., *History of the Arabs*, London, Macmillan, 1956.

Hudson, H.V., *The Great Divide* (Karachi: Oxford University Press), 1969.

Hussain, Azim, *Fazl-i-Husain* (Bombay: Longmans), 1946.

Hussain Riaz, *Iqbal Poet and His Politics* (New Delhi: Uppal Publishing House), 1987.

Iqbal Afzal (ed.), *Select Writings and Speeches of Maulana Mohamed Ali* (Lahore: Ashraf), 1944.

Ikram, S.M., *Modern Muslim India and the Birth of Pakistan* (Lahore: S.M. Ikram), 1965.

Iqbal, Mohammad, *Asrar-i-Khudi*, Lahore, Ghulam Ali, 1957 (Translated) as '*Secrets of Self*', New Delhi, Arnold-Heinemann, 1978.

Iqbal, Mohammad, *Asrar-i-Ramuz*, Lahore, Ghulam Ali, 1948.

Iqbal, Mohammad, *Bang-i-Dara*, Lahore, Ghulam Ali, 1953.

Iqbal, Mohammad, *Darb-i-Kalim*, Lahore, Ghulam Ali, 1959.

Iqbal, Mohammad, *Javid-Namah*, Lahore, Gulam Ali, 1932.

Iqbal, Mohammad, *Payam-i-Mashriq*, Lahore, Ghulam Ali, 1958.

Iqbal, Mohammad, *Ramuz-i-Bekhudi*, Lahore, Gulam Ali, 1918.

Iqbal Mohammed, *Six Lectures on the Reconstruction of Religious Thought in Islam*, Lahore, M. Ashraf, 1944.

Jafri, Ali Sardar & K.S. Duggal (eds.), *Iqbal Commemorative Volume*, New Delhi: All India Iqbal Centenary Celabration Committee, 1980.

Jain, M.S., *The Aligarh Movement: Its Origin and Development 1858-1906* (Agra: Sri Ram Mehra), 1965.

Jawed, Ajeet, *Secular and Nationalist Jinnah* (New Delhi: Kitab Publishing House), 1998.

Jinnah, Fatima, *My Brother* (Karachi: Quaid-e-Azam Academy), 1987.

Jinnah, M.A., *History of the Origin of Fourteen Points* (Bombay: n.d), mf (Australia: Flinders University).

Jinnah, M.A., *Speeches and Writings* (Lahore: Mustaffa), 1960.

Khalifah, Abdul Hakim, *Islam and Communism*, Lahore, Institute of Islamic Culture, 1953.

Khaliquzzaman, Chaudhri, *Pathway to Pakistan* (Lahore: Pakistan Longman), 1961.

Khan, Wali, *Facts are Facts—The Untold Story of Partisan* (New Delhi: Vikas Publications), 1987.

Khundmiri, S. Alam, "A Study of the Concepts of Transformation, Leadership and Freedom in the Political Philosophy of Iqbal" in Jafri Ali Sardar & K.S. Duggal (eds.), *Iqbal Commemorative Volume*, New Delhi, All India Iqbal Centenary Celebration Committee, 1980.

Khurshid, K.H., *Memories of Jinnah* (Karachi: Oxford University Press), 1990.

Kiernan, V.G., *Poems from Iqbal*, Bombay, Kutub, 1947.

Lentizion, Nehemia (ed.), *Conversion to Islam* (New York: Holmes and Meir), 1979.

Letters of Iqbal to Jinnah, published by Muhammad Ashraf quoted in Rafiq Zakaria, *The Man who Divided Pakistan* (Mumbai: Popular Prakashan), 2002.

Malik, Hafeez (ed.), *Pakistan Founders' Aspirations and To-days Realities* (New York: Oxford University Press), 2001.

Malik, Hafiz (ed.), *Iqbal Poet Philosopher of Pakistan* (New York: Columbia University Press), 1971.

Mansrgh, Nicholas, E.W.R. Lumby and Penderel Moon, The *Transfer of Power 1942-47* (London: Her Majesty Stationery Office), 1970-83.

McGrath, Allen, *The Destruction of Democracy in Pakistan* (Karachi: Oxford University Press), 1972.

Merriam, Allen Hayes, *Gandhi vs. Jinnah The Debate Over the Pakistan of India* (Calcutta: Minerva), 1980.

Mirza, Humanyun, *From Plassey to Pakistan* (New York: University Press of America), 1999.

Mohan, Sulakshana, *Pakistan Under Musharraf* (New Delhi: Indian Publishers), 1999.

Moon, Penderel, *Divide and Quit* (London: Chatto & Windus), 1964.

Morley, John Viscount, *Recollections*, Vol. II (London: Macmillan), 1918.

Mustafa, Zubeida (ed.), *The South Asian Century: 1900-1999* (London: Oxford University Press), 2000.

Naidu, Sarojini (ed.), *Mohammad Ali Jinnah, An Ambassador of Unity; His Speeches and Writings* (Madras: Ganesh), 1918.

Nanda, B.R., Gandhi, *Pan-Islamism, Imperialism and Nationalism in India* (Bombay: Oxford University Press), 1989.

Nehru, Jawaharlal, *An Autobiography* (London: The Bodley Head), 1937.

Niyazi, Sayyid Nazir, *Maktubat-i-Iqbal*, Karachi, Iqbal Academy, 1957.
Page, D., *Prelude to Pakistan* (Delhi: Oxford University Press), 1982.
Prashad, Rajendra, *India Divided* (Bombay: Kitabs), 1947.
Pirzada, S.S. (ed.), *Foundations of Pakistan, All India Muslim League Documents 1906-1947* (New Delhi: Metropolitan Book Company), 1982.
Pirzada, S.S. (ed.), *The Collected Works of Quaid-e-Azam Mohammad Ali Jinnah*, Vol. I (Karachi: East West Publishing), 1969.
Pirzada, Syed Sharifuddin, *Quaid-e-Azam Jinnah's Correspondence* (New Delhi: Metropolitan Book Company), 1981.
Pirzada, Syed Sharifuddin, *Some Aspects of Qaid e-Azam's Life* (Islamabad: National Commission on Historical and Cultural Research), 1978.
Ralhan, O.P. (ed.), *Hindu Mahasabha*, Vol. II (New Delhi: Anmol), 1997.
Ramani, Hashoo, *Kewa! Pakistan X-Rayed* (Delhi: New Age Publication), 1955.
Rawlinson, H.G., *The British Achievements in India* (London: William Hodge), 1948.
Resolutions of the All India Muslim League from March 1941 to April 1942 (Delhi: All India Muslim League).
Robinson, F., *Separation Among Muslims* (Cambridge: Cambridge University Press), 1974.
Roy, M.N., *Men I Met* (New Delhi: Lalvani Publications), 1968.
Sahni, J.N., *The Lid Off* (Bombay: Allied Publishers), 1971.
Salik, Abdul Majid, *Dhiker-i-Iqbal*, Lahore, Bazam-i-Iqbal, 1955.
Sayeed, K.B. *Pakistan The Formative Phase 1857-1948* (London: Oxford University Press), 1968.
Sayed, Khalid, B., *Pakistan: The Formation Phase* (London: Oxford University Press), 1968.
Sayed, Zaifullah and Samuel Baid, *Pakistan: An End Without a Beginning* (Delhi: Lancer International), 1985.
Sayed, K.B., *Pakistan: The Formative Phase 1857-1948* (London: Oxford University Press), 1968.
Sayid, M.H., *The Sound of Fury: A Political Study of Mohammad Ali Jinnah* (New Delhi: Akbar Publishing House), 1983.
Seervai, H.M., *Partition of India - Legend and Reality* (Bombay: Emmenem Publishers), 1989.
Shamloo, Latif Mohammad Sherwani, *Speeches and Statements of Iqbal* (Lahore: Al-Minor Academy), 1945.
Shamloo, Latif Ahmed Sherwani, *Speeches and Statements of Iqbal* (Lahore: Al-Minar Academy), 1995.
Sharif, Al-Mujahid, *Qaide-e-Azam* (Delhi: Low Price Publications), 1993.
Sharma, Jai Narain, "Irreligious Jinnah Wanted A Religious Pak", *The Tribune*, Chandigarh, 6 July 1997.

Sharma, Jai Narain, *Power, Politics and Corruption: A Gandhian Solution* (New Delhi: Deep and Deep), 2004.

Sharma, M.S.M., *Peeps Into Pakistan* (Patna: Pustak Bhandar), 1954.

Sharma, Rashmi (ed.), *Sonia* versus *Vajpayee* (New Delhi: Deep and Deep), 2004.

Sherwani, Latif Ahmad, *Pakistan Resolution to Pakistan* (Delhi: Daya Publishing House), 1985.

Sikander Hayat, *Pakistan Resolution Revisited* (Islamabad: National Institute of Historical and Cultural Research), 1999.

Singh, Anita Inder, *The Origin of the Partition of India 1936-47* (Bombay: Oxford University Press), 1987.

Singh, Iqbal, *The Ardent Pilgrim*, London, Longmans Green, 1951.

Sitaramayya, Pattabhi, *History of Indian National Congress* (Delhi: S. Chand & Co.), 1969.

Spear, Percival, *India Pakistan and the West* (London: Oxford University Press), 1958.

Sri Prakash, *Pakistan: Birth and Early Days* (New Delhi: Meenakshi Publications), 1956.

Sri Prakash, *Pakistan: Birth and Early Years* (Delhi: Meenakshi Prakashan), 1965.

Suda, J.P., *Main Currents of Social and Political Thought in India*, Vol. III (Meerut: K. Nath & Co.), 1973.

Symonds, Richard, *Making of Pakistan* (London: Faber and Faber), 1950.

Tara Chand, *History of Freedom Movement in India* (New Delhi: Publications Division, Ministry of Information and Broadcasting, Government of India), 1962.

Tendulkar, D.G, *Mahatma*, 8 Vols. (New Delhi: Publications Division, Ministry of Information and Broadcasting, Government of India), 1958.

The Concise Encyclopedia of Western Philosophy and Philosophers (London: Hutchinson), 1960.

The Encyclopedia of Social Sciences, Vol. III (London: Macmillan), 1957.

Vahid, S.A., *Iqbal His Art and Thought* (Hyderabad: Deccan), 1944.

Vahid, S.A., *Thoughts and Reflections of Iqbal* (Lahore: Mohammad Ashraf), 1964.

Varma, V.P., *Modern Indian Political Thought* (Agra: Laxmi Narayan Aggarwal), 1961.

Varma, V.P., *Modern Indian Political Thinker* (Agra: Laxmi Narayan Aggarwal), 1961.

Wahid-u-Din, Syed Faqir, *Ruzgar-i-Faqir*, Karachi Iqbal Academy, 1952.

Wells, Ian Bryant, *Jinnah's Early Politics Ambassador of Hindu Muslim Unity* (Delhi: Permanent Black), 2005.

Wolpert, Stanley, *Jinnah of Pakistan* (New York: Oxford University Press), 1984.

Index